Recent Advances in Monte Carlo Methods

Edited by Abdo Abou Jaoudé

Published in London, United Kingdom

Recent Advances in Monte Carlo Methods
http://dx.doi.org/10.5772/intechopen.1000269
Edited by Abdo Abou Jaoudé

Contributors
Abdujabar Rasulov, Adrian Northey, Barat Barati, David Lambert, Fariba Farhadi Birgani, Ian M. Robinson, Jan W. Dash, Mansour Zabihzadeh, Mark Flanagan, Masumeh Kaydani, Mohammad Javad Tahmasebi, Nahid Chegeni, Trevor C. Lipscombe, Tzu-Chien Hsueh, Xipei Yang

First published in London, United Kingdom, 2024 by IntechOpen
IntechOpen is the global imprint of INTECHOPEN LIMITED, registered in England and Wales, registration number: 11086078, 5 Princes Gate Court, London, SW7 2QJ, United Kingdom

British Library Cataloguing-in-Publication Data
A catalogue record for this book is available from the British Library

Additional hard and PDF copies can be obtained from orders@intechopen.com

Recent Advances in Monte Carlo Methods
Edited by Abdo Abou Jaoudé
p. cm.
Print ISBN 978-1-83769-362-7
Online ISBN 978-1-83769-361-0
eBook (PDF) ISBN 978-1-83769-363-4

For EU product safety concerns:
IN TECH d.o.o., Prolaz Marije Krucifikse Kozulić 3, 51000 Rijeka, Croatia,
info@intechopen.com or visit our website at intechopen.com.

Meet the editor

Abdo Abou-Jaoudé has been teaching for many years and has a passion for researching mathematics. He is currently an Associate Professor of Mathematics and Statistics at Notre Dame University-Louaizé (NDU), Lebanon. He holds a BSc and an MSc in Computer Science from NDU, and three Ph.D.s in Applied Mathematics, Computer Science, and Applied Statistics and Probability, all from Bircham International University, Spain. He also holds two Ph.D.s in Mathematics and Prognostics from the Lebanese University, Lebanon, and Aix-Marseille University, France. Dr. Abou-Jaoudé's broad research interests are in pure and applied mathematics. He has published twenty-three international journal articles and six contributions to conference proceedings, in addition to eleven books on prognostics, mathematics, and computer science.

Contents

Preface

It gives me great pleasure to introduce as well as to discuss, learn, solve, teach, and work with Monte Carlo Methods. In this book, *Recent Advances in Monte Carlo Methods*, we discuss some fundamental aspects of the theory of these Monte Carlo methods and explore their use to solve a large array of problems.

Monte Carlo methods, or Monte Carlo experiments, are a broad class of computational algorithms that rely on repeated random sampling to obtain numerical results. The underlying concept is to use randomness to solve problems that might be deterministic in principle. They are often used in physical and mathematical problems and are most useful when it is difficult or impossible to use other approaches. Monte Carlo methods are mainly used in three problem classes: optimization, numerical integration, and generating draws from a probability distribution.

Furthermore, in physics-related problems, Monte Carlo methods are useful for simulating systems with many coupled degrees of freedom, such as fluids, disordered materials, strongly coupled solids, and cellular structures. Other examples include modeling phenomena with significant uncertainty in inputs such as the calculation of risk in business and, in mathematics, the evaluation of multidimensional definite integrals with complicated boundary conditions. In application to systems engineering problems (space, oil exploration, aircraft design, etc.), Monte Carlo–based predictions of failure, cost overruns, and schedule overruns are routinely better than human intuition or alternative "soft" methods. In principle, Monte Carlo methods can be used to solve any problem having a probabilistic interpretation.

This volume illustrates the use of Monte Carlo Methods when applied to solve specific problems. It delves deeply into this exciting, profound, and modern field of mathematics and knowledge. The book discusses some fundamental aspects of applied Monte Carlo Techniques and, as such, it is a useful resource for scholars, researchers, and undergraduate and graduate students in pure and applied mathematics, physical sciences, engineering and technology, computer science, numerical analysis, scientific computing, and science in general.

Abdo Abou Jaoudé, Ph.D.
Notre Dame University-Louaizé,
Zouk Mosbeh, Lebanon

Chapter 1

Lanchester's Fighting Strength as a Battle Outcome Predictor Applied to a Simple Fire and Manoeuvre Wargame

Mark Flanagan, David Lambert, Trevor C. Lipscombe, Adrian Northey and Ian M. Robinson

Abstract

Lanchester's equations developed a mathematical understanding of the process of combat, leading to the concept of 'fighting strength', the product of fighting efficiency and numbers of troops squared. In this paper we demonstrate that 'fighting strength' is a key predictor of outcomes using a simple fire and manoeuvre wargame, set on a Mobius Strip. Lanchester's equations are solved showing 'saddle points', where beneath defeat is certain and above which victory is certain. The influence of tactics is explored using experimental design. The probability of loss in the game with consecutive dice rolls is solved. 'Fighting strength' predicted the final result in 33 out of 34 wargames with asymmetric forces. In addition Lanchester's equations also provide solutions for the % number of casualties in the wargames and the length of time each battle was fought. Based on initial pre-combat fighting efficiencies and numbers of troops between two opponents, a table of likely military strategies are presented to account for the differing 'fighting strengths' that best describe possible strategies that can succeed.

Keywords: Lanchester's equations, fighting strength, probability, wargame, move and fire

1. Introduction

> *I returned, and saw under the sun, that the race is not to the swift, nor the battle to the strong, neither yet bread to the wise, nor yet riches to men of understanding, nor yet favour to men of skill; but time and chance happeneth to them all. Ecclesiastes 9:11.*

Ecclesiastes tells us that life, including the outcome of battles, has a stochastic element. The study of military history tells us, in addition, that the numbers of troops engaged, their quality, weapons and tactics, together with their morale significantly influences the outcome. In 1916, F.W. Lanchester [1] developed a mathematical understanding of the process of combat, and this is reviewed in Appendix A. The loss of active fighting troops with fire exchange follows a differential equation, whose solution leads to the concept of 'fighting strength' in a square law relationship.

$$\text{'Fighting strength'} = P\,N^2. \tag{1}$$

where P is the 'efficiency' or probability of hitting an opponent in a fire exchange, removing them from further combat, and N is the number of troops available to fight in the fire exchange. The simultaneous and continuous fire exchange in the battle leads to a steady reduction in numbers of combatants for each side with the outcome driven by the fighting strength. In the example below unless the Blue force has a significantly higher probability P_B of hitting an opponent in a fire exchange then the disparity in numbers compared to the Red force will eventually dominate (**Figure 1**).

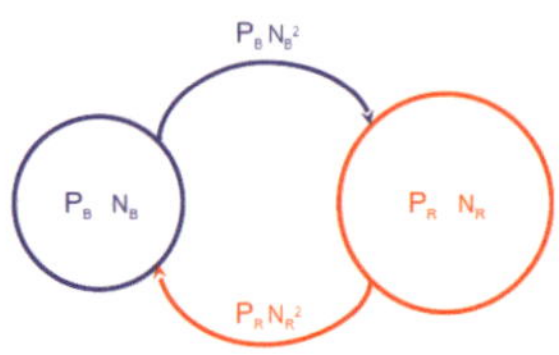

Figure 1.
Schematic representation of the Lanchester model for combat for Red and Blue forces, with the area of the circle proportional to the number of troops N_B (Blue) & N_R (Red) in each force, and the probability of hitting an opponent P_B (Blue) and P_R (Red).

Lanchester's equations suggest that the fighting strength of two opposing forces indicate the outcome of the battle before it takes place: the side with the greater fighting strength being able to eliminate its rival should it stand and fight to the end. Based on this, it is possible to devise a table of strategies [2] that looks at the best options for a Red force attacking a Blue force, by looking carefully at variations in combat efficiency and numbers and how these affect fighting strength.

The suggested battle strategies for a Red force fighting a Blue force based on Lanchester's fighting strength include (**Table 1**).

	Lower efficiency ($P_R < P_B$) than Blue	Same efficiency ($P_R = P_B$) as Blue	Better efficiency ($P_R > P_B$) than Blue
Smaller numbers than Blue $N_R < N_B$	Fabian strategy [5] $P_R N_R^2 \ll P_B N_B^2$	Fabian strategy [5], or tactics of division if unable to avoid battle [4]	Tactics of concentration [2]
Same numbers as Blue $N_R = N_B$	Fabian strategy [5], or tactics of division if unable to avoid battle [4]	Tactics of division [4]	Tactics of concentration [2]
Higher numbers than Blue $N_R > N_B$	Tactics of numbers [3]	Tactics of numbers [3]	Vernichtungsschlacht [1] $P_R N_R^2 \gg P_B N_B^2$

Table 1.
Classification of possible tactics for a Red force fighting a Blue force with underlying variations to fighting strength.

1. *Vernichtungsschlacht—Battle of Annihilation*. Move immediately to optimal firing range and destroy the Blue force, since $P_R N_R^2 \gg P_B N_B^2$.

2. Tactics of concentration. Divide the Blue forces into smaller groups, so that the Red force has greater fighting strength in each battle against the smaller groups of Blue forces, destroying them in turn.

3. Tactics of numbers. Red forces can defeat a smaller Blue force by preventing them from dividing. Red forces (i.e. prevent Blue forces from using tactics of concentration against them), engaging at optimal firing range.

4. Tactics of division. Set up small battles more favourable to Red forces, in an attempt to reduce down the Blue force numbers. This could involve decoys, traps and surprise manoeuvres such as tunnels.

5. Fabian strategy—avoid pitched battles. Manoeuvre, engaging with Blue forces sparingly on best terms of engagement as possible in a war of attrition, since $P_R N_R^2 \ll P_B N_B^2$.

Later we will explore the influence of such tactics on the outcome of a simple wargame.

Lanchester's equations have been applied with success to many different historical battles including naval (Trafalgar [3–5], Dogger Bank [6], Jutland [7], Midway [8–10]); land (American Civil War [11, 12]), WW2 (Kursk [13–16], Iwo Jima [17], Insurgency [18–22]); and air (WW2 Battle of Britain [23]). They have also been applied to social insect warfare [24, 25]. Additional papers that examine or extend the theory are listed in Appendix A. In this paper, we test the fighting-strength concept by introducing a simple manoeuvre-and-fire wargame, one that readily becomes a *Vernichtungsschlacht*—Battle of Annihilation—because of the playing surface.

2. Simple manoeuvre and fire wargame—Möbius Mayhem

The map represents a space station that is a Möbius strip, a continuous featureless surface from which escape is not possible and which players representing Red and Blue forces of space marines seek to control. These forces are armed with direct fire weapons, with unlimited ammunition and whose probability of hitting an opponent increases as the range decreases (**Figure 2**).

The coloured and lettered hexes are 'trapdoors', which allow units to move up or down the surface to the corresponding coloured and lettered hex, which is normally 20 hexes vertically distant and with a mirrored translation in symmetry left to right, or vice versa. If a unit moves off the top of the map, it reappears at the bottom and vice versa since the surfaces join together. Likewise, if it moves off one of the sides, it reappears on the other side with a 20 hex vertically distant mirrored transition like the trapdoors. The switches in symmetry and the continuous nature of the playing surface is a function of being a Möbius strip [26] (without the mirror symmetry, the space would be the surface of a torus).

Players use a fixed turn sequence (IgoYougo). The sequence of play is Move, then Fire, based on the system described in [27]. Units that move cannot fire. They may move up to 5 hexes per turn (0 to 5 Movement Points, MP, allocation), and may pass through the same hex as another unit, but cannot remain in the same hex as another unit at the end of the turn. Going through a trapdoor and reappearing on the board elsewhere costs 1MP, and units cannot loiter in a trapdoor. Trapdoors are 'jump cuts' on the Möbius surface—ultrafast 'tunnels' to be used as players see fit. Units that fire cannot move, and a clear line of sight is needed to fire. They can fire from the top of the board to the bottom and vice versa since the surfaces join, but cannot fire across the sides, or through trapdoors. To hit and eliminate an enemy with regular quality troops, count the distance in hexes, and roll a 1D6. The score must match or exceed the range in hexes, with six hexes maximum range. Thus, at a range of 6 hexes, you need a roll of 6. Troop quality can also vary. Poor troop quality has a (−1) modifier applied to all 1D6 throws. Good troop quality has a (+1) modifier applied to all 1D6 throws. This affects the probability of hitting and eliminating an enemy.

There is no morale effect in the game, and given no retreat is possible from the playing surface with no terrain blocking, the game is played as a *Vernichtungsschlacht*—Battle of Annihilation—until one side is eliminated.

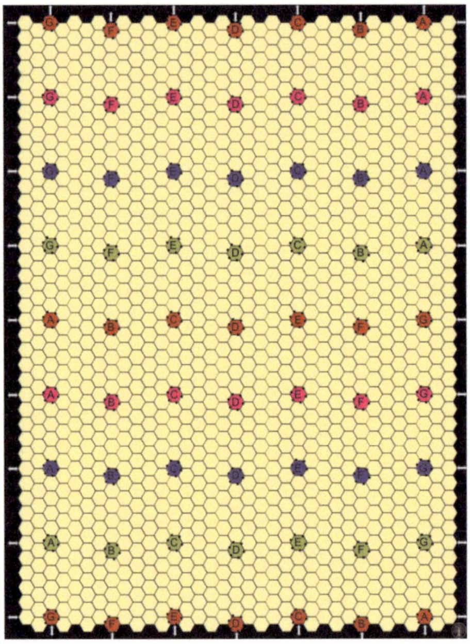

Figure 2.
Playing surface for the Möbius Mayhem wargame.

The game is available in an online playable form [27]. It includes a supporting spreadsheet with inbuilt dice rollers that rapidly calculate combat outcomes. It also tracks casualties, provides a solution to Lanchester equations, and suggests scripted strategies on a random basis. In addition to being playable online (via Google drive or similar method), it also includes a pdf of the map that can be printed onto A0 paper with each hex being 25 mm or 1″ in size for table-top play, combined with suitable figures of choice. The game relies on having unbiased methods to generate 1D6 rolls. These include rolling trusted physical dice; random number generation by spreadsheet with suitable algorithms; Google Dice; or using an absolute random number sequence such as pi, which can be readily converted to 1D6 as suggested in [28].

Since the game plays as a *Vernichtungsschlacht*, it is an ideal laboratory in which to test Lanchester's equations, since the combat probabilities are well-defined. It has aspects of chess play, namely an opening (advance to contact), leading to a middle game where exchange of pieces occurs, and finally an endgame where the winning side closes down the losing side.

1D6 modifier	1 hex	2 hex	3 hex	4 hex	5 hex	6 hex
(+1)	1	1	5/6	2/3	1/2	1/3
0	1	5/6	2/3	1/2	1/3	1/6
(−1)	5/6	2/3	1/2	1/3	1/6	0

Table 2.
Probability, P, vs. range in hexes to secure a hit and remove of an enemy unit for troops of good quality (+1), regular quality (0) and poor quality (−1).

Table 3 highlights the assumptions in Lanchester's theory and the Möbius Mayhem wargame (MMW) helps highlight their differences.

Variable	Lanchester's model	Möbius Mayhem wargame (MMW)
Combat efficiency, P	Fixed	Changes with range and troop quality
Movement	Not allowed	Not obligatory, in a wargame that has sequential turns between forces. Units which move cannot fire in a turn
Fire exchange	Deterministic and continuous	Not obligatory, and stochastic, in a wargame that has sequential turns between forces. Units may be out of range to receive or give fire Units that fire cannot move in a turn
Fire allocation	Not allowed, assumed area firing	Allowed within the combat range

Table 3.
Differences between Lanchester's model and MMW.

Lanchester's model of combat is based upon simultaneous and continuous exchange of fire. In real combat, one side may often open fire before the other when it comes into a suitable range. This can confer an advantage if it achieves a large reduction in opposing force numbers before receiving counter fire [29]. In the game, the play is fixed turn sequence, which results in a sequential exchange of fire as shown below (**Figure 3**).

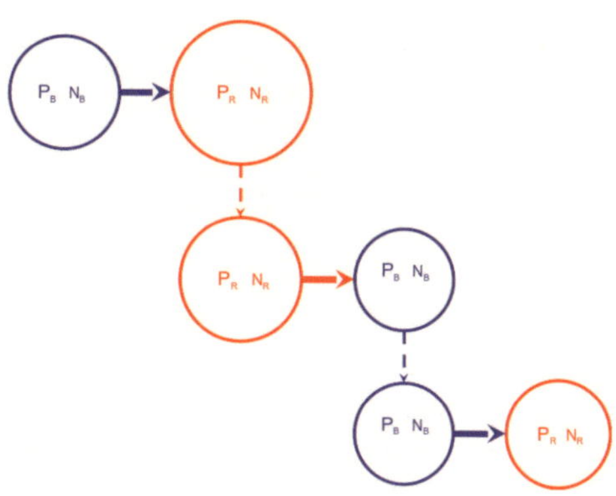

Figure 3.
Schematic representation of the Lanchester model for combat as a stochastic Markov chain, where fire exchange is sequential rather than simultaneous. The area of each circle is proportional to the number of troops N_B & N_R in each force. Note the steady reduction in numbers of troops for both Red and Blue forces due to attrition in combat, which is driven in part by stochastic dice rolls against fixed probabilities of hitting an opponent.

This can be described as a stochastic Markov Chain, where the next step in the game, in terms of combat outcome, depends on the state attained in the previous event with respect to fighting strength. To this extent, the fighting strength constantly changes turn by turn as numbers decline and ranges alter, which adjusts the probability of hitting an opponent. Consequently, players need to pay constant attention to the strategies for combat suggested in **Table 1** as the game proceeds. Despite the differences highlighted above, MMW serves as a means of studying whether such a simple wargame follows aspects of the Lanchester model in an accessible form. Lanchester models are deterministic, and there is value in exploring deterministic models.

3. Example games to test Lanchester's equations using MMW

3.1 Fighting strength vs. different troop levels, quality and combat range

Based on the probability outcomes for success in combat in the game with ranges as defined in **Table 2**, we can examine the fighting strength for a series of different troop levels (10, 12, 14, 16, and 20 units) and quality (poor, regular, good). These are tabulated below.

Number of troops (−1 on dice roll)	1 hex	2 hexes	3 hexes	4 hexes	5 hexes	6 hexes
10 (−1)	83.3	66.7	50.0	33.3	16.7	0.0
12 (−1)	120.0	96.0	72.0	48.0	24.0	0.0
14 (−1)	163.3	130.7	98.0	65.3	32.7	0.0
16 (−1)	213.3	170.7	128.0	85.3	42.7	0.0
20 (−1)	333.3	266.7	200.0	133.3	66.7	0.0

Table 4.
*Fighting strength ($P N^2$) from probabilities of hitting vs. range in hexes and number of available poor quality troops. (−1) is added to all 1D6 casts, so the bottom row in **Table 2** is used to calculate the fighting strength.*

Number of troops (+1 on dice roll)	1 hex	2 hexes	3 hexes	4 hexes	5 hexes	6 hexes
10 (+1)	100.0	100.0	83.3	66.7	50.0	33.3
12 (+1)	144.0	144.0	120.0	96.0	72.0	48.0
14 (+1)	196.0	196.0	163.3	130.7	98.0	65.3
16 (+1)	256.0	256.0	213.3	170.7	128.0	85.3
20 (+1)	400.0	400.0	333.3	266.7	200.0	133.3

Table 5.
*Fighting strength ($P N^2$) from probabilities of hitting vs. range in hexes and number of available good quality troops. (+1) is added to all 1D6 casts, so the top row in **Table 2** is used to calculate the fighting strength.*

Number of troops (0 on dice roll)	1 hex	2 hexes	3 hexes	4 hexes	5 hexes	6 hexes
10 (0)	100.0	83.3	66.7	50.0	33.3	16.7
12 (0)	144.0	120.0	96.0	72.0	48.0	24.0

Number of troops (0 on dice roll)	1 hex	2 hexes	3 hexes	4 hexes	5 hexes	6 hexes
14 (0)	196.0	163.3	130.7	98.0	65.3	32.7
16 (0)	256.0	213.3	170.7	128.0	85.3	42.7
20 (0)	400.0	333.3	266.7	200.0	133.3	66.7

Table 6.
Fighting strength ($P N^2$) from probabilities of hitting vs. range in hexes and number of available regular quality troops. (0) is added to all 1D6 casts, so the middle row in ***Table*** *2 is used to calculate the fighting strength.*

Furthermore, closing the range has a large effect in improving fighting strength for both an attacker and a defender. Given the game plays with a fixed turn sequence (IgoYougo), an attacker must move into range before firing, necessarily exposing themselves to fire exchange before returning fire in the next turn. In this paper for all battles Red is the attacker advancing on Blue defending, generally resulting in Blue firing first from the turn sequence. This may give the Blue defender an inbuilt advantage and is a tactical consequence of the simple game rules not envisaged in Lanchester's original work, which has continuous simultaneous firing. Nevertheless, MMW can examine pre-contact fighting strength (determined at the longest mutual range for fire exchange) and its ability to predict the winner in a battle. This is done in a series of asymmetrical games, which have different numbers of troops and troop quality to test whether values of pre-contact fighting strength are robust indicators of success or failure. Following this an examination of symmetrical pre-contact fighting strengths with troops of identical levels and quality is conducted, together with near symmetrical fighting strengths. Lanchester [1] stated in battles for equal fighting strength

In the case of equal forces the two conjugate curves become coincident; there is a single curve of logarithmic form; the battle is prolonged indefinitely. Since the forces actually consist of a finite number of finite units (instead of an infinite number of infinitesimal units), the end of the curve must show discontinuity, and break off abruptly when the last man is reached; the law based on averages evidently does not hold rigidly when the numbers become small. Beyond this, the condition of two equal curves is unstable, and any advantage secured by either side will tend to augment.

Lanchester's equations indicate that forces with matched fighting strengths will eliminate each other, the consequence of an exchange of fire that is both deterministic and continuous, with a proviso that attrition may make one side gain numerical advantage towards the end of the combat. MMW has an exchange of fire that is sequential and stochastic, inevitably resulting in asymmetric numbers of forces evolving during the battle. If the forces have a matched probability of hitting in fire exchange then a difference in fighting strength can only arise from a reduction in numbers. Since the fighting strengths are matched and Lanchester's model gives mutual reduction in forces under these conditions, two explanations in this wargame for eventual defeat by one side are offered.

a. For losses between forces initially stochastically matched, a tipping point will be likely reached when the number of units for one side reduces to 5, as seen from **Figures** 7 and 8 in Appendix B. Any further fall in unit numbers from 5 causes the probability of hitting to change by about 10% which is a large deleterious effect. If a force has greater than 5 units and the opponent falls to 5 or less, then it becomes increasingly unlikely that the disparity in forces can be reversed in the game. A number differential of at least 6 to 5 gives a difference of $(6/5)^2 = 44\%$ in fighting strengths, given the dependency on the square of the numbers concerned, indicating a likely win for the force with more units.

b. Alternatively the side that inflicts the first major losses on their opponent gains an advantage in fighting strength, and is likely to secure victory in the remainder of the fire exchange. This advantage can arise stochastically.

3.2 Asymmetric fighting strength games

In this section a test is made of fighting strength and whether it can predict the outcome of a battle ahead of first contact and subsequent fire exchange. Fighting efficiency is altered by troop quality at the longest range mutual fire exchange can occur and combining this with the number of troops this alters the fighting strength. Section 3.2.1 looks at a Red force that outnumbers a Blue force by 2:1 but with a disparity in troop quality, in a test of whether successive waves of attacks or a mass attack is more likely to win. Section 3.2.2 examines in a systematic manner using a designed-experiment approach the fighting strength based on fighting efficiency and numbers for a Red force employing varying tactics against a defensive Blue force with fixed numbers. This determines whether fighting strength is the key determinant, or whether tactics are also important. Section 3.2.3 uses a randomised sequence of fighting strengths by altering troop quality and numbers for both the Red and Blue forces, with both employing a steady firing line, again as a check of the importance of fighting strength as a battle outcome indicator. In all these battles a prediction of the eventual winner from the initial fighting strength and the % survivors predicted from Eq. (A11) is made and compared to the outcome.

3.2.1 Test of concentration of forces: single attack vs. waves

Three runs were conducted to demonstrate Lanchester's concept of concentration of force for effective fighting. A force of 21 poor quality Red troops attack a force of defending 10 good quality Blue troops. In the first run 3 waves of 7 Red troops attack, spaced by 2 moves apart. In the second run 2 waves of 11 and 10 Red troops attack, spaced by 2 moves apart. In the third run a single wave of 21 Red troops attack. The results are below and shown graphically in [30] as animated gifs, together with a plot of % remaining units and fighting strengths per turn. The data in runs 1a-3, **Table 12**, Appendix C highlight the importance of attacking in strength, with the Red force defeated in runs 1 & 2, but overwhelming Blue in run 3 despite having the same number of Red units available in all runs. The battle prediction follows the expected pattern from the initial fighting strengths calculated, and in all further runs in this paper the Red force attacks in one wave.

3.2.2 Designed experiment

A previous study used a factorial designed experiment [31] on a simple wargame called 'Take that Hill' [32]. The study showed the advantage of this method in exploring games quickly and systematically. In this paper, 5 variables are examined using another type of experimental design [33] to test the effect of numbers vs. simple tactics for a Red force in an attacking role against a fixed number of Blue forces in a defensive role. The variables are set at low (−1) and high (+1) values. They include Red Force Tactics [Fire and manoeuvre, closing to (range) (+1) or Hit at (range) then retire regroup repeat (−1)], Red Force Attack Range {(max range in hexes − 2 (+1) or max range in hexes (−1)}, Red Force Attack Method {attacking nearest enemy regardless of fighting strength (+1), or attacking flank or weakest group (−1)], Red Force Trapdoor use {using trapdoors (+1) or not using trapdoors (−1)}, and finally Red Force Numbers (lower (−1) or higher (+1)}. In the case of fire and manoeuvre, the initial contact troops will begin firing at the enemy, then subsequent troops will join this firing line, allowing the surviving initial troops to rush forward by one hex and so on until the optimal firing distance by the attack range is reached. Previous work [34] indicated an effect on survivability in manoeuvre-and-fire combat to rush distance for an attacking force, so the value chosen in this game is representative of the rush distance suggested. The experimental pattern is shown in **Table 7**.

These can be used to produce scripted strategies for the Red force. In each game the initial fighting strengths, expected asymptote survival %, result, final survival %, and the game time in moves is tracked for analysis.

RunOrder	Red Force Tactics A	Red Force Attack Range B	Red Force Attack Method C	Red Force Trapdoors D	Red Force Numbers E
1	1	1	1	−1	−1
2	1	−1	−1	1	−1
3	−1	−1	−1	−1	−1
4	−1	1	−1	1	1
5	1	1	−1	−1	1
6	1	−1	1	1	1
7	−1	1	1	1	−1
8	−1	−1	1	−1	1

Table 7.
Designed experiment using 8 runs for 5 variables, labelled A–E (see Appendix B).

3.2.2.1 Red force with regular quality and 12 or 16 units vs. Blue force with good quality and 10 units

The matching numbers for a Red force of regular quality achieving balance in fighting strength from Eq. (A12) in Appendix A is 14.1 units against 10 Blue units of good quality. Careful selection of the Red force numbers around this saddle point of 14.1 units suggests 12 units for a low level and 16 units for a high level. The two values for Red force numbers straddle this value and so serve as a critical test of the influence of fighting strengths. The designed experiment in **Table 7** then generates the following scripted attacking strategies for the Red force (**Table 8**).

Run	Scripted attacking Red force strategies
4	Manoeuvre and fire, closing to max range—2 attacking nearest group not using trapdoors with 12 units
5	Manoeuvre and fire, closing to max range attacking flank or weakest group using trapdoors with 12 units

Run	Scripted attacking Red force strategies
6	Hit at max range attacking flank or weakest group then retire, regroup and repeat not using trapdoors with 12 units
7	Hit at max range—2 attacking flank or weakest group then retire, regroup and repeat using trapdoors with 16 units
8	Manoeuvre and fire, closing to max range—2 attacking flank or weakest group not using trapdoors with 16 units
9	Manoeuvre and fire, closing to max range attacking nearest group using trapdoors with 16 units
10	Hit at max range—2 attacking nearest group then retire, regroup and repeat using trapdoors with 12 units
11	Hit at max range attacking nearest group then retire, regroup and repeat not using trapdoors with 16 units

Table 8.
Scripted attacking Red force strategies in the trial shown in Table 13 , Appendix C.

Number of troops (dice modifer)	1 hex	2 hexes	3 hexes	4 hexes	5 hexes	6 hexes
10 (+1)	100.0	100.0	83.3	66.7	50.0	33.3
12 (0)	144.0	120.0	96.0	72.0	48.0	24.0
16 (0)	256.0	213.3	170.7	128.0	85.3	42.7

Table 9.
Fighting strength vs. combat range for forces in the trial.

From the probabilities in **Tables 4** and **5** the following relationships of fighting strength for an initial Blue force of 10 units, of good quality vs. two Red forces of 12 & 16, of regular quality are shown in **Table 9**.

Note that at the longest attack range of 6 hexes, a Blue force of 10 units is stronger than a Red force of 12 units, since $P_R N_R^2$ (24.0) < $P_B N_B^2$ (33.3). This advantage drops at an attack range of 4 hexes, when $P_R N_R^2$ (72.0) > $P_B N_B^2$ (66.7). This illustrates the effect of probability from combat range on fighting strength. It is clearly in Blue forces interest to keep Red force at arms length, but for a Red force to "Grab Their Belts to Fight Them!" and close the distance with the Blue force, accepting their losses on the way in. Consequently we might expect from fighting strength a force of 10 Blue units of good quality to beat a regular force of 12 Red units of regular quality at long range, but lose at shorter range, other things being equal. Also note that at the longest attack range of 6 hexes, a force of 10 Blue units at good quality is weaker than a force of 16 Red units at regular quality, since $P_R N_R^2$ (42.7) > $P_B N_B^2$ (33.3). This advantage increases at an attack range of 4 hexes, when $P_R N_R^2$ (128.0) $\gg$ $P_B N_B^2$ (66.7). Overall, we might anticipate a Blue force of 10 units of good quality can beat a Red force of 12 units of regular quality, but lose to a Red force of 16 units, also of regular quality. The games were played and all are available as animated gifs [30], together with plots of % remaining units and fighting strength turn by turn. The results are given in runs 4-11, **Table 13**, Appendix C.

In 8 out of 8 battles the initial fighting strength indicator based on long range combat correctly predicted the outcome of the battle, suggesting the identified saddle point in Red numbers by Eq. (A12) is real. Detailed analysis by best subsets regression analysis of each variable on the % survivors for each side, and number of moves until the end of the wargame is described in Appendix C. The most significant variable to explain the eventual outcome of each battle is the number of Red forces involved (E). In addition, all tactical effects (A–D) do influence the time for the battle to end in annihilation of one side.

3.2.2.2 Red force with poor quality and 16 or 20 units vs. Blue force with good quality and 10 units

The matching numbers for a Red force of poor quality achieving balance in fighting strength from Eq. (A12) is 17.3 units against 10 Blue units of good quality. Careful selection of the Red force numbers around the saddle point of 17.3 units suggests 16 units for a low level and 20 units for a high level. The two values for Red force numbers straddle this and so serve as a critical test of the fighting strengths.

The designed experiment in **Table 7** generates the following scripted attacking strategies for the Red force (**Table 10**).

Run	Scripted attacking Red force strategies
12	Manoeuvre and fire, closing to max range—2 attacking nearest group not using trapdoors with 16 units
13	Manoeuvre and fire, closing to max range attacking flank or weakest group using trapdoors with 16 units
14	Hit at max range attacking flank or weakest group then retire, regroup and repeat not using trapdoors with 16 units
15	Hit at max range—2 attacking flank or weakest group then retire, regroup and repeat using trapdoors with 20 units
16	Manoeuvre and fire, closing to max range—2 attacking flank or weakest group not using trapdoors with 20 units
17	Manoeuvre and fire, closing to max range attacking nearest group using trapdoors with 20 units
18	Hit at max range—2 attacking nearest group then retire, regroup and repeat using trapdoors with 16 units
19	Hit at max range attacking nearest group then retire, regroup and repeat not using trapdoors with 20 units

Table 10.
Scripted attacking Red force strategies in the trial shown in Table 14, Appendix C.

Number of troops (dice modifer)	1 hex	2 hex	3 hex	4 hex	5 hex	6 hex
10 (+1)	100.0	100.0	83.3	66.7	50.0	33.3
16 (−1)	213.3	170.7	128.0	85.3	42.7	0.0
20 (−1)	333.3	266.7	200.0	133.3	66.7	0.0

Table 11.
Fighting strength vs. combat range for forces in the trial.

We note from **Tables 4** and **6** the following relationships for an initial Blue force of 10, of good quality (+1 on 1D6 dice rolls) vs. two Red forces of 16 & 20, of poor quality (−1 on 1D6 dice rolls) (**Table 11**).

Note that at the longest attack range of 6 hexes, the Blue force is stronger than all Red forces, since Red cannot hit. At an attack range of 5 hexes, a force of 10 Blue units is stronger than a force of 16 Red units, since $P_R N_R^2$ (42.7) ≪ $P_B N_B^2$ (50.0). This advantage changes at an attack range of 4 hexes, when $P_R N_R^2$ (85.3) ≫ $P_B N_B^2$ (66.7). At the longest mutual attack range of 5 hexes, a force of 10 Blue units is weaker than a force of 20 Red units with poor quality, since $P_R N_R^2$ (66.7) ≫ $P_B N_B^2$ (50.0). This advantage to Red increases at an attack range of 4 hexes, when $P_R N_R^2$ (133.3) ≫ $P_B N_B^2$ (66.7). Consequently we might anticipate a Blue force of 10 units of good quality can beat a poor quality Red force of 16 units, but lose to a Red force of 20 units, unless they are able to keep the Red forces at a maximum firing range of 6 hexes.

The wargames were played and are available as animated gifs [30], together with the % surviving units and the fighting strengths, turn by turn. The outcomes are shown in runs 12-19 **Table 14**, Appendix C.

In 7 out of 8 battles the initial fighting strengths correctly predicted the outcome of the battle, again suggesting the identified saddle point in Red numbers by Eq. (A12) is real. Run 8 was incorrectly predicted. 10 Blue units fired on 20 Red units causing 5 hits at odds of hitting of 1/2. In return, 15 Red units fired but obtained no hits. At odds of success of 1/6, 15/6 = 2.5 hits might be *deterministically* expected, which would have reduced the Blue units significantly. However from Appendix B Eq. (A17), the odds of not hitting with 15 units each with a separate dice roll is 6.5%. Since this did not occur a second fire exchange from Blue caused a further 5 hits on Red. At this point the fighting strengths for 10 Red (16.7) vs. 10 Blue (50.0) indicated a Blue win, the eventual outcome. The predicted results based on initial fighting strengths are not deterministic, and this demonstration of the stochastic fire exchange in the game gives a probability of the estimated eventual outcome at 95% confidence for asymmetric battles.

Detailed analysis by best subsets regression analysis of the variables on the % survivors for each side, and number of moves until the end of the wargame is described in Appendix C. The most significant variable to explain the eventual outcome of each battle is the number of Red forces involved (E). In addition, tactical effects (A–D) do influence number of survivors and the time for the battle to end in annihilation of one side, especially variables A and B.

From the battles in Section 3.2, small differences in numbers combined with changes in probabilities produces different fighting strengths, and different end battle outcomes. Runs 7–9 and 11 had Red with 16 regular troops and a fighting strength of 42.3 against Blue with 10 good troops and a fighting strength of 33.3. In each battle the Red force annihilated the Blue force. Likewise, runs 12–14 and 18 had a Red force with 16 poor troops and a fighting strength of 42.7 against a Blue force with 10 good troops and a fighting strength of 50.0. In each battle the Blue force annihilated the Red force. Clearly the fighting strength appears to be a good indicator of success.

3.2.3 Randomised trials of asymmetric Red and Blue forces using steady firing line at max range

Fighting strengths listed in **Tables 4** to **6** were studied by randomising troop numbers and quality for both Red and Blue forces. Each side was free to use their own tactics. This produced runs 20-34 in **Table 15**, Appendix C, together with their outcomes. This produced the following trial, together with their outcomes.

Across all the asymmetric battles in runs 1–34, the outcome was correctly predicted using the pre-contact fighting strength indicator for 33 out of 34 battles or

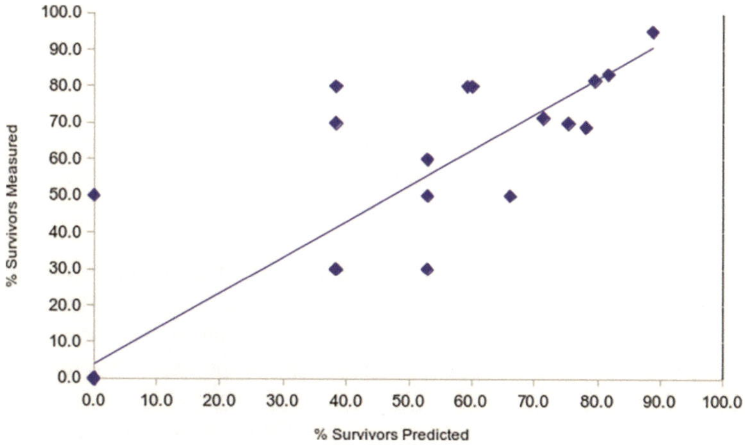

Figure 4.
*% survivors predicted from Eq. (A11) vs. % survivors measured for Blue force from all runs from data in **Tables 12** to **15**. The slope, intercept and correlation coefficient for the data are m = 0.98, c = 4.08 and R = 0.91 (32 DoF) respectively.*

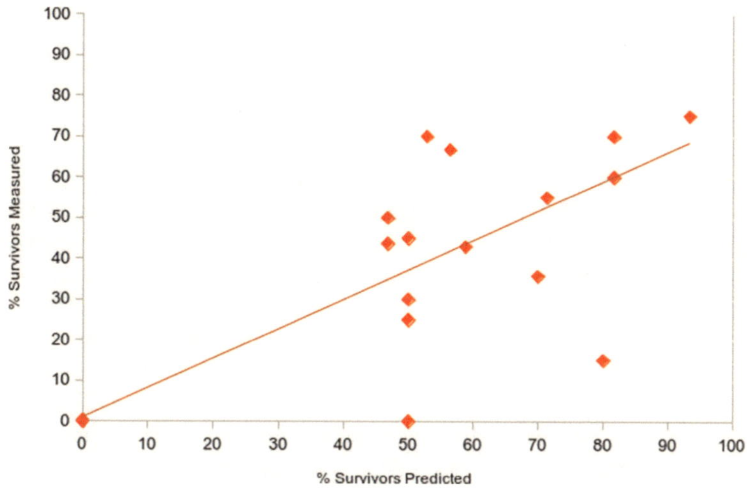

Figure 5.
*% survivors predicted from Eq. (A11) vs. % survivors measured for Red force from all runs from data in **Tables 12** to **15**. The slope, intercept and correlation coefficient for the data are m = 0.72, c = 1.09 and R = 0.86 (32 DoF).*

97%. Plotting the % survivors predicted from Eq. (A11) vs. % survivors measured is also instructive using the data in **Tables 12** to **15**.

The critical value of Pearson's Correlation Coefficient at 99% confidence and 32 degrees of freedom (DoF) is 0.449. The two values for the correlation coefficient in **Figures 4** and **5** exceed this, so the trends are significant at the 99% confidence level, implying the predictions in the game of winning or losing outcomes (by fighting strength) and the % survivors (by Lanchester's Eq. (A11)) are reliable, probably at the 95% level, and the asymmetric battles in the game conform well to Lanchester's expectations.

3.3 Symmetrical fighting strength battles

3.3.1 Symmetrical fighting strength battles

Section 3.2 indicates that fighting strength is a good indicator for success in MMW. Real battles are rarely conducted between forces that exactly match each other in fighting strength. Hobby games tend to have symmetrically matched starting conditions in a sense of fairness between the players.

Runs [35–49] were fought between Red and Blue with matched fighting strengths based on equal troop quality to test the concepts of (a & b) in Section 3.1, with Red launching a central attack using a steady firing line against Blue in a trial of strength. This produced runs 35-49 given in **Table 16**, Appendix C, together with their outcomes.

The progress in each battle clearly shows stochastic behaviour in the fire exchange, and that explanation (a) in Section 3.1 appears to apply. Red won 6 of the battles with Blue winning 9 of the battles. This may indicate a slight bias towards the Blue defending force, especially when one considers the % remaining units left at the end of the battles. By this measure, when Red won their battles they had fewer % remaining units compared to the battles that Blue won. This gives some credence for explanation (b) in Section 3.1 also being present in the games.

3.3.2 Nearly symmetrical forces with closely matched fighting strengths

Saddle points where the fighting strengths between different numbers and troop qualities between Red and Blue have already been identified from Eq. (A12) and used in the asymmetric battles. In this section, values close to these were chosen for the Red forces producing 3 battles with nearly symmetrical fighting strengths. There is a slight imbalance in forces which from Lanchester's equations should lead to the Blue Force winning each battle. This produced runs 50-52 given in **Table 17**, Appendix C, together with their outcomes.

Red won 1 of the battles with Blue winning 2 of the battles. Overall from the symmetric and nearly symmetric battles in Sections 3.3.1 and 3.3.2 there are 7 Red wins (avg % units remaining 38.5%), 11 Blue wins (avg % units remaining 70.2%). There is some evidence for defender bias (b) in these battles. There are also plenty of examples for exchange of fire leading to condition (a) in the runs. Stochastic behaviour in fire exchange leads to the balance of forces swinging to one side, and thereafter, Lanchester's insights apply.

Taylor [35] introduced the concept of 'fighting intensity', $\sqrt{P_R * P_B}$ where P_B and P_B are the probabilities of hitting an opponent. Taking the probabilities for mutual fire exchange for Red and Blue from **Table 2**, calculating $\sqrt{P_R * P_B}$ and plotting against the number of moves for the battle to end for each of the runs in **Tables 12–16** gives

The critical value of Pearson's Correlation Coefficient at 99% confidence and 47 degrees of freedom (DoF) is 0.372, suggesting the trend is significant at the 99% confidence level. 'fighting intensity' is a guide for battle duration in MMW, with higher 'fighting intensity' i.e. improved odds in hitting reducing the number of moves to complete the battle (**Figure 6**).

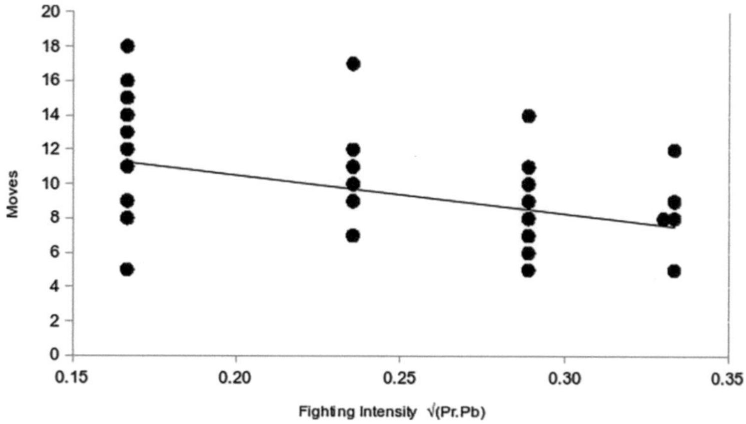

Figure 6.
Moves for battle to end vs. 'fighting intensity' from all runs in ***Tables** 3 and **12–16**. The correlation coefficient for the data is R = −0.441 (47 DoF).*

4. Conclusions

When is a battle lost? von Clausewitz [36] gives his own definition.

> *"The result of the whole combat consists in the sum total of the results of all partial combats; but these results of separate combats are settled by different considerations. First by the pure moral power in the mind of the leading officers. Secondly, by the quicker melting away of our troops, which can be easily estimated in the slow and relatively little tumultuary course of our battles. Thirdly, by lost ground.*

Perhaps in battles that can be described by Lanchester's equations the result is mostly a foregone conclusion if the Fighting Strengths are widely dissimilar before battle is joined. However *Vernichtungsschlacht* remains a Fata Morgana of the modern military experience [37].

There is a clear mismatch between the assumptions in Lanchester's model and the simple manoeuvre and fire MMW, which uses sequential turns and stochastic exchange of fire. Despite these differences it appears that Lanchester's concept of initial fighting strength PN^2 has some validity in predicting the outcomes in the study, with 33 out of 34 battles successfully predicted resulting in the *Vernichtungsschlacht* outcome. This suggests that the concept behind maximal fighting strength, namely a concentration of forces and their use in fire exchange to best advantage, firing first if possible, are key to success in MMW, as in real combat.

The battle outcomes from initial fighting strength at the longest mutual combat range predicted the final result 33 out of 34 times across the asymmetric runs. This indicates it is a good prognosis for eventual battle outcome. Note that neither explanation of (a) being reduced to 5 units or (b) inflicting first major losses appeared to apply for almost all the asymmetric battles, where victory went to the side at 95% confidence with fighting strength advantage, regardless of being Red or Blue, with the exception of run 19. In the symmetric games, once attrition swings the force balance to one side the fighting strength in combat is also a good prognosis for eventual battle outcome.

Military reality is more complex than either Lanchester's model or a simple manoeuvre and fire wargame. Regarding Lanchester's model, in addition to the variables in **Table 11** the following are also neglected. The model considers homogeneous forces only. No replacements or manoeuvre to withdraw are possible. There is no consideration of non combat losses (e.g. surrenders, desertions). Suppressive effects of weapons rather than outright elimination are not considered. There are no logistical considerations. Effects of terrain or weather conditions are not considered. There are numerous academic papers which try to adapt Lanchester's equations to reflect the subtlety of real combat, and the caveats with the model with some success which are listed in the references [35, 38–45] for interested readers. Lanchester equations have been applied with success, primarily to idealised situations, given that they do not—nor do they try to—take into account fatigue, morale, or in-battle decision-making. The aim of a professional wargame is to address variables such as the above and include as many as desired in the rules and fighting maps used. It remains to be seen if the concept of fighting strength can be used to predict the outcomes of more sophisticated wargames.

Whether a simple or professional wargame, as in war itself, the Biblical warning from Luke still stands.

> *Or what king, going to make war against another king, sitteth not down first, and consulteth whether he be able with ten thousand to meet him that cometh against him with twenty thousand? Luke 14:31.*

Appendix

A. Lanchester's equations

Quoting directly from Lanchester's original work [1].

> *"If, again, we assume equal individual fighting value, and the combatants otherwise on terms of equality, each man will in a given time score, on an average, a certain number of hits that are effective; consequently, the number of men knocked out per unit time will be directly proportional to the numerical strength of the opposing force."*

We now prove this statement. Suppose we have two forces in combat, Red and Blue, who allocate fire to a new target when a previous one is hit and eliminated from further action. The original numbers of Red and Blue forces are N_R and N_B. The 'efficiency' or probability of hitting an opponent in a fire exchange, removing them from further combat, is P_R for Red and P_B for Blue forces.

If we assume simultaneous and continuous exchange of fire the rate of Red forces losses is

$$\frac{dN_R}{dt} = -P_B N_B \tag{A1}$$

and the rate of Blue force losses is

$$\frac{dN_B}{dt} = -P_R N_R \tag{A2}$$

Define the fighting strength F by the equation:

$$PN^2 \tag{A3}$$

and consider the difference, S, in the fighting strengths between the two forces:

$$S = P_R N_R^2 - P_B N_B^2 \tag{A4}$$

Differentiate this with respect to time, to obtain:

$$\frac{dS}{dt} = 2\left[P_R N_R \frac{dN_R}{dt} - P_B N_B \frac{dN_B}{dt}\right] \tag{A5}$$

Now substitute in the time derivatives in [A2] and [A3] to see:

$$\frac{dS}{dt} = 2\left[P_R N_R \frac{dN_R}{dt} - P_B N_B \frac{dN_B}{dt}\right] = 2[P_R N_R(-P_B N_B) + P_B N_B P_R N_R] = 0 \tag{A6}$$

This means that the difference in the fighting strengths between the two forces does not change during the battle. Mathematically, this means:

$$P_R N_R^2(t) - P_B N_B^2(t) = constant \tag{A7}$$

Apply this formal at two times. The first, at the opening of the conflict and time t = 0, the second at the end of the conflict, when t = T. We assume that at the end, the Blue forces have been annihilated, which will occur if $P_R N_R^2 > P_B N_B^2$. As a consequence:

$$P_R N_R^2(T) = P_R N_R^2(0) - P_B N_B^2(0) \tag{A8}$$

Divide through by P_R to obtain:

$$N_R^2(T) = N_R^2(0) - \frac{P_B}{P_R} N_B^2(0) \tag{A9}$$

So that:

$$N_R^2(T) = N_R^2(0)\left[1 - \frac{P_B N_B^2(0)}{P_R N_R^2(0)}\right] \tag{A10}$$

Or equivalently:

$$N_R(T) = N_R(0)\sqrt{1 - \frac{P_B N_B^2(0)}{P_R N_R^2(0)}} \tag{A11}$$

Consequently, the number of Red forces remaining at the end of the conflict depends on both the initial number of Red forces, but also on the ratio of the fighting strengths of the Blue to Red forces. The greater the fighting strength of the Blue forces, the more losses the Red forces will suffer.

If the initial fighting strength for the Red Force is less than the Blue Force, then $P_R N_R^2(0) < P_B N_B^2(0)$, and the Red force will lose the battle and be annihilated.

Eq. (A11) will predict the number of Blue survivors under these conditions if we interchange the Red and Blue subscripts.

In the case of 'symmetrical' matched fighting strengths at the start of fire exchange between Red and Blue forces $P_R N_R^2(0) = P_B N_B^2(0)$. Examining this gives the adjustments needed to the starting 'asymmetrical' unmatched fighting strengths to make them 'symmetrical' matched fighting strengths:

$$\text{For initial force numbers } N_R(0) = N_B\sqrt{\left(\frac{P_B}{P_R}\right)} \tag{A12}$$

$$\text{For effectiveness } P_R(0) = N_B(0)\left(\frac{P_B}{P_R}\right)^2 \tag{A13}$$

Eqs. (A12) and (A13) predict the numbers for the Blue force if we interchange the Red and Blue subscripts.

Taylor [35] gives a solution for the time to annihilation in Lanchester's model, but this does not apply in this study due to the different time dependence of sequential play of the wargame used. Further details on the theory and extensions therein can be found in the following papers [38–45].

B. Probability of rolling a target value in repeated 1D6 casts

Define the number of faces, f, on a 1D6 that yields the desired outcome for a hit from rolling 1D6 n times. The chances of rolling the required value, f, in repeated n casts of 1D6 is

$$1 - \left(\frac{(6-f)}{6}\right)^n \tag{A14}$$

Likewise the chances of not rolling the required value, f, in repeated n casts of 1D6 is

$$\left(\frac{(6-f)}{6}\right)^n \tag{A15}$$

By example the chance of rolling a six (f = 1) in 15 casts is

$$1 - \left(\frac{(6-1)}{6}\right)^{15} = 0.935 \tag{A16}$$

And the chance of not rolling a six (f = 1) in 15 casts is

$$\left(\frac{(6-1)}{6}\right)^n = 0.065 \tag{A17}$$

In the game, the probability needed to hit changes with quality and range as shown in **Table 2**. For regular quality units at six hexes, a 6 is required (f = 1) to hit. At 5 hexes a cast of 5 or 6 is required (f = 2) to hit. At 4 hexes a cast of 4, 5 or 6 is required (f = 3) to hit. Taking (A14) and (A15) and plotting them against 20 1D6 rolls vs. target scores (2–6 to 6) gives

Note the shape of change in probability with number of casts and target value for a hit to occur. As hits accrue to a force the number of units that can fire reduces which translates to moving along the LHS of each curve. Changing the range translates to jumping between curves.

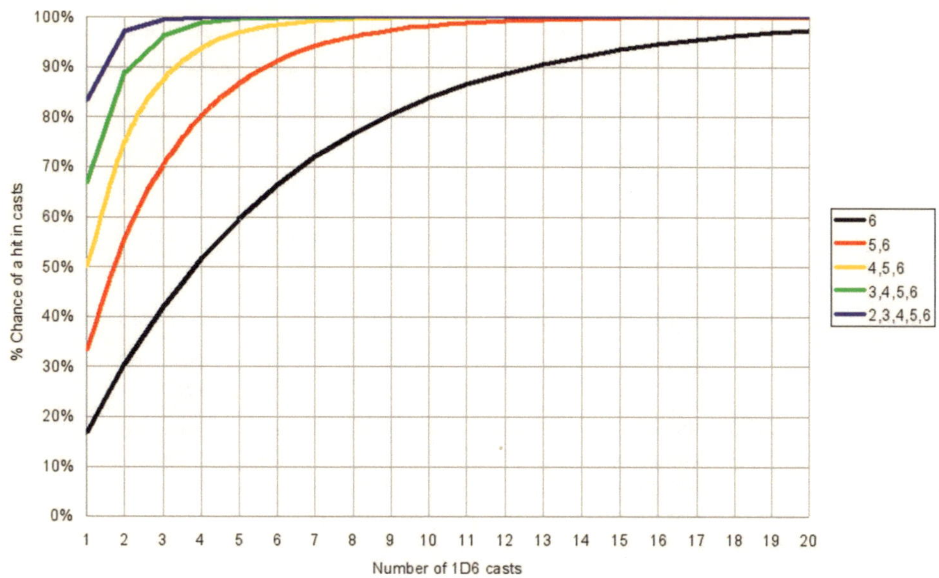

Figure 7.
Chance of a single hit from increasing 1D6 casts from Eq. (A14) vs. target scores.

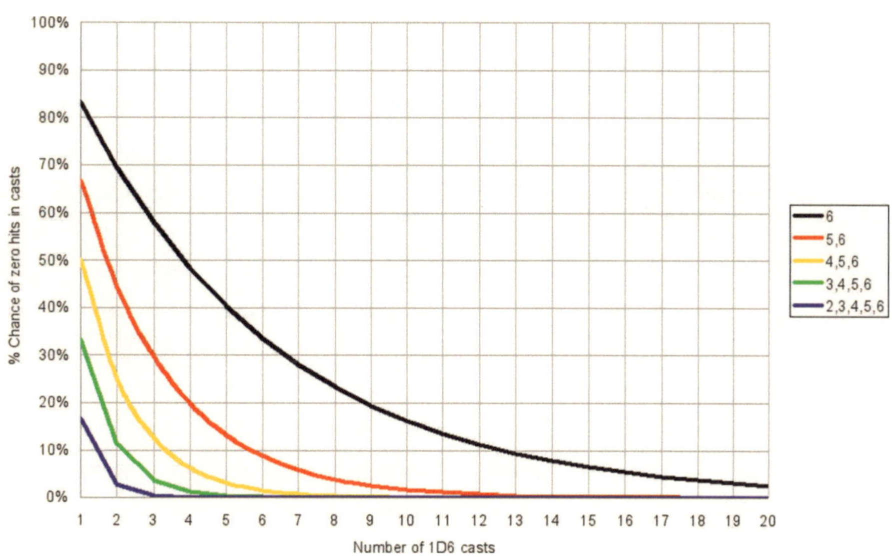

Figure 8.
Chance of not achieving a single hit from increasing 1D6 casts from Eq. (A15) vs. target scores.

C. Table of results: predicted vs measured battle outcomes to test concentration of forces using Lanchester's Equations

Run	Red	Blue	Fighting strengths (Red to Blue)	Prediction	% Survivors predicted (Red to Blue)	Outcome	% Survivors measured (Red to Blue)	Moves
1a	7 (−1)	10 (+1)	8.2 < 50.0	Blue win	0.0% & 91.5%	Blue win	0.0% & 80.0%	—
1b	7 (−1)	10 (+1)	8.2 < 50.0	Blue win	0.0% & 82.3%	Blue win	0.0% & 70.0%	—
1c	7 (−1)	10 (+1)	8.2 < 50.0	Blue win	0.0% & 75.3%	Blue win	0% & 70.0%	11
2a	11 (−1)	10 (+1)	20.2 < 50.0	Blue win	0.0% & 77.2%	Blue win	0.0% & 80.0%	—
2b	10 (−1)	10 (+1)	16.7 < 50.0	Blue win	0.0% & 62.9%	Blue win	0.0% & 80.0%	9
3	21 (−1)	10 (+1)	73.3 > 50.0	Red win	56.5% & 0.0%	Red win	66.7% & 0.0%	8

Table 12.
Predicted vs. measured battle outcomes to test concentration of forces in an asymmetric battle between Blue and Red attacking in waves.

Run	Red	Blue	Fighting strengths (Red to Blue)	Prediction	% Survivors predicted (Red to Blue)	Outcome	% Survivors measured (Red to Blue)	Moves
4	12 (0)	10 (+1)	24.0 < 33.3	Blue win	0.0% & 52.9%	Blue win	0.0% & 50.0%	7
5	12 (0)	10 (+1)	24.0 < 33.3	Blue win	0.0% & 52.9%	Blue win	0.0% & 30.0%	12
6	12 (0)	10 (+1)	24.0 < 33.3	Blue win	0.0% & 52.9%	Blue win	0.0% & 60.0%	17
7	16 (0)	10 (+1)	42.3 > 33.3	Red win	46.8% & 0.0%	Red win	50.0% & 0.0%	10
8	16 (0)	10 (+1)	42.3 > 33.3	Red win	46.8% & 0.0%	Red win	50.0% & 0.0%	9
9	16 (0)	10 (+1)	42.3 > 33.3	Red win	46.8% & 0.0%	Red win	50.0% & 0.0%	7
10	12 (0)	10 (+1)	24.0 < 33.3	Blue win	0.0% & 52.9%	Blue win	0.0% & 60.0%	9
11	16 (0)	10 (+1)	42.3 > 33.3	Red win	46.8% & 0.0%	Red win	43.7% & 0.0%	11

Table 13.
Predicted vs. measured battle outcomes for the scripted strategy runs for a Blue force of 10 units of good quality vs. two Red forces of 12 & 16, of regular quality. The fighting strength assumes contact at 6 hexes, i.e. maximum mutual range.

Run	Red	Blue	Fighting strengths (Red to Blue)	Prediction	% Survivors predicted (Red to Blue)	Outcome	% Survivors measured (Red to Blue)	Moves
12	16 (−1)	10 (+1)	42.7 < 50.0	Blue win	0.0% & 38.3%	Blue win	0.0% & 70.0%	7
13	16 (−1)	10 (+1)	42.7 < 50.0	Blue win	0.0% & 38.3%	Blue win	0.0% & 30.0%	10
14	16 (−1)	10 (+1)	42.7 < 50.0	Blue win	0.0% & 38.3%	Blue win	0.0% & 80.0%	8
15	20 (−1)	10 (+1)	66.7 > 50.0	Red win	50.0% & 0.0%	Red win	25.0% & 0.0%	8
16	20 (−1)	10 (+1)	66.7 > 50.0	Red win	50.0% & 0.0%	Red win	45.0% & 0.0%	10

Run	Red	Blue	Fighting strengths (Red to Blue)	Prediction	% Survivors predicted (Red to Blue)	Outcome	% Survivors measured (Red to Blue)	Moves
17	20 (−1)	10 (+1)	66.7 > 50.0	Red win	50.0% & 0.0%	Red win	30.0% & 0.0%	9
18	16 (−1)	10 (+1)	42.7 < 50.0	Blue win	0.0% & 38.3%	Blue win	0.0% & 30.0%	7
19	20 (−1)	10 (+1)	66.7 > 50.0	Red win	50.0% & 0.0%	Blue win	0.0% & 50.0%	14

Table 14.
Predicted vs. measured battle outcomes for the scripted strategy runs for a Blue force of 10 units of good quality vs. two Red forces of 16 & 20, of poor quality. The fighting strength assumes contact at 5 hexes, i.e. maximum mutual range.

Run	Red	Blue	Fighting strengths (Red to Blue)	Prediction	% Survivors predicted (Red to Blue)	Outcome	% Survivors measured (Red to Blue)	Moves
20	12 (+1)	16 (+1)	48.0 < 85.3	Blue win	0.0% & 66.1%	Blue win	0.0% & 50.0%	8
21	14 (−1)	12 (+1)	32.7 < 72.0	Blue win	0.0% & 81.6%	Blue win	0.0% & 83.3%	7
22	12 (−1)	14 (0)	24.0 < 65.3	Blue win	0.0% & 79.5%	Blue win	0.0% & 81.7%	7
23	10 (+1)	10 (−1)	50.0 > 16.7	Red win	81.6% & 0.0%	Red win	60.0% & 0.0%	6
24	20 (+1)	20 (−1)	200.0 > 66.7	Red win	81.6% & 0.0%	Red win	70.0% & 0.0%	6
25	16 (0)	20 (0)	42.7 < 66.7	Blue win	0.0% & 60.0%	Blue win	0.0% & 80.0%	13
26	10 (0)	12 (−1)	33.3 > 24.0	Red win	52.9% & 0.0%	Red win	70.0% & 0.0%	8
27	14 (+1)	16 (0)	65.3 > 42.7	Red win	58.9% & 0.0%	Red win	42.9% & 0.0%	11
28	16 (−1)	20 (+1)	42.7 < 200.0	Blue win	0.0% & 88.7%	Blue win	0.0% & 95.0%	5
29	20 (−1)	14 (−1)	66.7 > 32.7	Red win	71.4% & 0.0%	Red win	55.0% & 0.0%	11
30	10 (−1)	16 (−1)	16.7 < 42.7	Blue win	0.0% & 78.1%	Blue win	0.0% & 68.8%	5
31	12 (0)	14 (+1)	24.0 < 65.3	Blue win	0.0% & 71.4%	Blue win	0.0% & 71.4%	7
32	16 (+1)	10 (+1)	78.1 > 33.3	Red win	93.3% & 0.0%	Red win	75.0% & 0.0%	12
33	20 (0)	12 (0)	66.7 > 24.0	Red win	80.0% & 0.0%	Red win	15.0% & 0.0%	18
34	14 (0)	10 (0)	32.7 > 16.7	Red win	70.0% & 0.0%	Red win	35.7% & 0.0%	16

Table 15.
Predicted vs. measured battle outcomes for the asymmetric battles between Blue and Red forces of 10–20 units with poor to good quality. The fighting strength assumes fire exchange at maximum mutual range.

Run	Red	Blue	Fighting strengths (Red to Blue)	Prediction	% Survivors predicted (Red to Blue)	Outcome	% Survivors measured (Red to Blue)	Moves
35	10 (−1)	10 (−1)	16.7 = 16.7	Draw	0.0% & 0.0%	Red win	40.0% & 0.0%	15
36	10 (0)	10 (0)	16.7 = 16.7	Draw	0.0% & 0.0%	Red win	30.0% & 0.0%	12
37	10 (+1)	10 (+1)	33.3 = 33.3	Draw	0.0% & 0.0%	Blue win	0.0% & 80.0%	8
38	12 (−1)	12 (−1)	24.0 = 24.0	Draw	0.0% & 0.0%	Red win	58.3% & 0.0%	8
39	12 (0)	12 (0)	24.0 = 24.0	Draw	0.0% & 0.0%	Blue win	0.0% & 75.0%	9

Run	Red	Blue	Fighting strengths (Red to Blue)	Prediction	% Survivors predicted (Red to Blue)	Outcome	% Survivors measured (Red to Blue)	Moves
40	12 (+1)	12 (+1)	48.0 = 48.0	Draw	0.0% & 0.0%	Blue win	0.0% & 91.7%	5
41	14 (−1)	14 (−1)	32.7 = 32.7	Draw	0.0% & 0.0%	Red win	57.1% & 0.0%	9
42	14 (0)	14 (0)	32.7 = 32.7	Draw	0.0% & 0.0%	Blue win	0.0% & 78.6%	9
43	14 (+1)	14 (+1)	65.3 = 65.3	Draw	0.0% & 0.0%	Red win	14.3% & 0.0%	9
44	16 (−1)	16 (−1)	42.7 = 42.7	Draw	0.0% & 0.0%	Red win	31.3% & 0.0%	14
45	16 (0)	16 (0)	42.7 = 42.7	Draw	0.0% & 0.0%	Blue win	0.0% & 62.5%	12
46	16 (+1)	16 (+1)	85.3 = 85.3	Draw	0.0% & 0.0%	Blue win	0.0% & 68.8%	5
47	20 (−1)	20 (−1)	66.7 = 66.7	Draw	0.0% & 0.0%	Blue win	0.0% & 45.0%	11
48	20 (0)	20 (0)	66.7 = 66.7	Draw	0.0% & 0.0%	Blue win	0.0% & 75.0%	12
49	20 (+1)	20 (+1)	133.3 = 133.3	Draw	0.0% & 0.0%	Blue win	0.0% & 55.0%	9

Table 16.
Predicted vs. measured battle outcomes for symmetric battles between Blue and Red forces of 10–20 units with regular quality. The fighting strength assumes contact at maximum mutual range depending on troop quality.

Run	Red	Blue	Fighting strengths (Red to Blue)	Prediction	% Survivors predicted (Red to Blue)	Outcome	% Survivors measured (Red to Blue)	Moves
50	7 (0)	10 (−1)	16.3 ≈ 16.7	Blue win	0.0% & 14.2%	Blue win	0.0% & 30.0%	10
51	14 (−1)	10 (0)	32.7 ≈ 33.0	Blue win	0.0% & 14.0%	Red win	42.9% & 0.0%	9
52	17 (−1)	10 (+1)	48.2 ≈ 50.0	Blue win	0.0% & 19.1%	Blue win	0.0% & 50.0%	8

Table 17.
Predicted vs. measured battle outcomes for nearly symmetrical battles between Blue and Red forces of 7–17 units with varying quality. The fighting strength assumes contact at maximum mutual range depending on troop quality.

D. Designed experiment analysis

The statistical methodology for analysing designed experiments is described in [30]. The designed experiment followed the pattern given in **Table 4**.

Best subsets regression analysis and regression equations were performed using Minitab v18 with +1, −1 for each variable shown in **Table 4**, together with the data in **Tables 8** and **17**. The variables were Red Force Tactics (A), Red Force Attack Range (B), Red Force Attack Method (C), Red Force using Trapdoors (D) and Red Force Numbers (E), with each variable set to −1 or +1 as described in **Table 4**. The responses studied were % Blue and Red remaining forces, and the time in moves for the wargame to end with the annihilation of one side.

D.1 Red force with regular quality and 12 or 16 units vs. Blue force with good quality and 10 units

Best subsets regression and regression equations identified were
See **Tables 18–20.**
i.e. only the number of Red troops mattered, and is independent of tactics.

Var	R-sq	Ad. R-sq	C-p	s	A	B	C	D	E
1	**89.3**	**87.5**	**0.8**	**10.0**					**X**
1	3.6	0.0	39.2	30.0	X				
2	92.9	90.0	1.2	8.9	X				X
2	90.2	86.2	2.4	10.5		X			X
3	93.7	89.1	2.8	9.4	X	X			X
3	93.7	89.1	2.8	9.4	X		X		X
4	94.6	87.5	4.4	10.0	X	X	X		X
4	94.6	87.5	4.4	10.0	X	X		X	X
5	95.5	84.4	6.0	11.2	X	X	X	X	X

Table 18.
Blue % remaining: best subsets regression (best fit in bold text).

Var	R-sq	Ad. R-sq	C-p	s	A	B	C	D	E
1	**99.4**	**99.3**	**2.0**	**2.2**					**X**
1	0.1	0.0	946.3	28.0				X	
2	99.5	99.3	3.0	2.2	X				X
2	99.5	99.3	3.0	2.2			X		X
3	99.6	99.3	4.0	2.2	X		X		X
3	99.6	99.3	4.0	2.2			X	X	X
4	99.7	99.3	5.0	2.2	X		X	X	X
4	99.7	99.3	5.0	2.2	X	X	X		X
5	99.8	99.3	6.0	2.2	X	X	X	X	X

Table 19.
Red % remaining: best subsets regression (best fit in bold text).

Var	R-sq	Ad. R-sq	C-p	s	A	B	C	D	E
1	33.3	22.2	192.0	2.9	X		X		
1	24.5	41.0	122.0	2.5		X			
2	57.8	41.0	122.0	2.5	X		X		
2	57.8	41.0	122.0	2.5		X	X		
3	82.3	69.0	52.0	1.8	X	X	X		

Var	R-sq	Ad. R-sq	C-p	s	A	B	C	D	E
3	68.7	45.2	92.0	2.4		X	X		X
4	93.2	84.1	22.0	1.3	X	X	X		X
4	88.4	73.0	36.0	1.7	X	X	X	X	
5	**99.3**	**97.6**	**6.0**	**0.5**	**X**	**X**	**X**	**X**	**X**

Table 20.
Moves until end of wargame: best subsets regression (best fit in bold text).

The regression equation is

$$\%\text{Blue remaining} = 25.0–25.0^{*}\text{E} \quad \text{(A18)}$$

i.e. only the number of Red troops mattered, and is independent of tactics.
The regression equation is

$$\%\text{Red remaining} = 24.2 + 24.2^{*}\text{E} \quad \text{(A19)}$$

i.e. all variables matter, and tactics plays a role.
The regression equation is

$$\text{Moves} = 10.3–1.50^{*}\text{A}–1.50^{*}\ \text{B}–1.75^{*}\ \text{C}–0.750^{*}\ \text{D}–1.00^{*}\ \text{E} \quad \text{(A20)}$$

All variables have an effect, including the tactical ones. The shortest battle duration will happen when all variables are set to high values i.e. Fire and manoeuvre, closing to (range defined by force attack), Force Attack Range at (max range in hexes —2), Red Force Attack Method, attacking nearest enemy regardless of fighting strength, Red Force Attack Method, attacking nearest enemy regardless of fighting strength, Red Force Trapdoor use, using trapdoors, and Red Force Numbers set to higher levels.

D.2 Red force with regular quality and 12 or 20 units vs. Blue force with good quality and 10 units

Best subsets regression and regression equations identified were
See **Tables 21–23.**
i.e. only variables A, B, D and E, the number of Red troops mattered, and tactics (A, B, D) do matter.
The regression equation is

$$\%\text{Blue remaining} = 32.5–7.50^{*}\text{A} - 7.50^{*}\text{B} - 17.5^{*}\text{D} - 20.0^{*}\text{E} \quad \text{(A21)}$$

i.e. only variables A, B, C and E the number of Red troops mattered, and tactics (A–C) do matter.
The regression equation is

$$\%\text{Red remaining} = 12.5 + 6.25^{*}\ \text{A} + 5.00^{*}\ \text{B} - 5.00^{*}\ \text{C} + 12.5^{*}\ \text{E} \quad \text{(A22)}$$

Var	R-sq	Ad. R-sq	C-p	s	A	B	C	D	E
1	44.8	5.5	15.8	25.7					X
1	34.3	23.3	19.5	27.9				X	
2	79.0	70.6	5.5	17.3				X	X
2	51.0	31.5	15.5	26.5	X				X
3	85.3	74.3	5.3	16.2		X		X	X
3	85.2	74.3	5.3	16.2	X		X		X
4	**91.6**	**80.4**	**5.0**	**14.1**	**X**	**X**		**X**	**X**
4	88.1	72.3	6.2	16.8	X		X	X	X
5	94.4	80.4	6.0	14.1	X	X	X	X	X

Table 21.
Blue % remaining: best subsets regression (best fit in bold text).

Var	R-sq	Ad. R-sq	C-p	s	A	B	C	D	E
1	54.3	46.7	2.5	13.23					X
1	13.6	0.0	0.0	18.2	X				
2	67.9	55.1	2.5	12.15	X				X
2	63	48.3	3.2	13.04			X		X
3	76.6	59.1	3.3	11.59	X	X			X
3	76.6	59.1	3.3	11.59	X	X	X		X
4	**85.3**	**65.8**	**4.1**	**10.61**	**X**	**X**	**X**		**X**
4	77.2	46.7	5.2	13.23	X		X	X	X
5	85.9	50.5	6	12.75	X	X	X	X	X

Table 22.
Red % remaining: best subsets regression (best fit in bold text).

Var	R-sq	Ad. R-sq	C-p	s	A	B	C	D	E
1	27.5	15.4	0.0	2.1					X
1	27.5	15.4	0.0	2.1		X			X
2	54.9	36.9	0.5	1.8		X			X
2	35.9	10.3	1.6	2.2				X	X
3	**63.4**	**35.9**	**2.0**	**1.8**		**X**		**X**	**X**
3	55.3	21.7	2.5	2.0	X	X			X
4	63.9	15.4	4.0	2.1		X	X	X	X
4	63.7	15.4	4.0	2.1	X	X		X	X
5	64.1	0.0	6.0	2.6	X	X		X	X

Table 23.
Moves until end of wargame: best subsets regression (best fit in bold text).

i.e. all variables matter, and tactics plays a role.

The regression equation is

$$\text{Moves} = 9.13–1.12^{*}\,\text{B}–0.625^{*}\ \text{D} + 1.13^{*}\ \text{E} \tag{A23}$$

Overall all variables matter, with the number of Red units the most significant. Tactics plays a role in time for a battle to conclude and also in the second designed experiment for % troops remaining. This highlights the weakness of tactical choice A at a negative value (hit at (range) then retire, regroup, and repeat) in the game. This exposes Red to Blue counterfire three times for every time it fires in return (once when closing to distance, once when counterfiring and once when Blue returns fire before retiring, then regrouping and repeating the process again).

Author details

Mark Flanagan[1], David Lambert[2], Trevor C. Lipscombe[3], Adrian Northey[2] and Ian M. Robinson[2*]

1 NHS Business Service Authority, Newcastle upon Tyne, UK

2 Hearts of Oak, North Yorks, UK

3 Catholic University of America, Washington, DC, USA

*Address all correspondence to: a.anser4u@gmail.com

References

[1] Lanchester FW. The principle of concentration: N- square law, Chapter 5. In: Aircraft in Warfare: The Dawn of the Fourth Arm. London: Constable; 1916. Available from: https://tinyurl.com/3h9j63xv

[2] Chan P. The Lanchester square law: Its implications for force structure and force preparation of Singapore's operationally-ready soldiers' pointer. Journal of Singapore Armed Forces;**42**(2):42-60. Available from: https://tinyurl.com/2k3sx868

[3] Kingman JFC. Stochastic aspects of Lanchester's theory of warfare. Journal of Applied Probability. 2002;**39**(3):455-465. Available from: http://www.jstor.com/stable/3216070

[4] McCue B. Lanchester and the Battle of Trafalgar. Phalanx. 1999;**32**(4):10-13, 22. Available from: https://www.jstor.org/stable/43962654

[5] Nash DH. Differential equations and the Battle of Trafalgar. The College Mathematics Journal. 1985;**16**(2):98-102. Available from: http://www.jstor.com/stable/2686209

[6] MacKay N, Price C, Wood J. Weighing the fog of war: Illustrating the power of Bayesian methods for historical analysis through the Battle of the Dogger Bank. Historical Methods. 2016;**49**(2):80-91. DOI: 10.1080/01615440.2015.1072071

[7] MacKay N, Price C, Wood J. Weight of Shell must tell: A Lanchestrian reappraisal of the Battle of Jutland. History. 2016;**101**(347):536-563. DOI: 10.1111/1468-229X.12241

[8] Taylor TC. A simple, functional model of modern naval conflict. Naval War College Review. 1995;**48**(3):99-112. Available from: http://www.jstor.com/stable/44642811

[9] Bongers A, Torres JL. Revisiting the Battle of midway. Military Operations Research. 2020;**25**(2):49-68. Available from: https://www.jstor.org/stable/10.2307/26917214

[10] Armstrong MJ. The salvo combat model with a sequential exchange of fire. The Journal of the Operational Research Society. 2014;**65**(10):1593-1601. Available from: http://www.jstor.com/stable/24505020

[11] Weiss HK. Combat models and historical data: The U.S. civil war. Operations Research. 1966;**14**(5):759-790. Available from: http://www.jstor.com/stable/168777

[12] Johnson RL. Lanchester's Square Law in Theory and Practice, AD-A225-484. Army Command and General Staff College Fort Leavenworth; 1989. Available from: https://apps.dtic.mil/sti/pdfs/ADA225484.pdf

[13] Lucas TW, Dinges JA. The effect of Battle circumstances on fitting Lanchester equations to the Battle of Kursk. Military Operations Research. 2004;**9**(2):17-30. Available from: http://www.jstor.com/stable/43940971

[14] Lucas TW. Fitting Lanchester equations to the battles of Kursk and Ardennes. Naval Research Logistics. 2004;**51**:95-116. DOI: 10.1002/nav.10101

[15] Speight LR. Within-campaign analysis: A statistical evaluation of the Battle of Kursk. Military Operations Research. 2011;**16**(2):41-62. Available from: http://www.jstor.com/stable/43941475

[16] Kuikka V. A combat equation derived from stochastic modeling of attrition data. Military Operations Research. 2015;**20**(3):49-69. Available from: https://www.jstor.org/stable/10.2307/24838621

[17] Engel JH. A verification of Lanchester's Law. Journal of the Operations Research Society of America. 1954;**2**(2):163-171. Available from: http://www.jstor.com/stable/166602

[18] Schaffer MB. Lanchester models of guerrilla engagements. Operations Research. 1968;**16**(3):457-488. Available from: http://www.jstor.com/stable/168576

[19] Kress M, Szechtman R. Why defeating insurgencies is hard: The effect of intelligence in counterinsurgency operations—A best-case scenario. Operations Research. 2009;**57**(3):578-585. Available from: http://www.jstor.com/stable/25614776

[20] Atkinson MP, Gutfraind A, Kress M. When do armed revolts succeed: Lessons from Lanchester theory. Journal of the Operational Research Society. 2012;**63**(10):1363-1373. Available from: http://www.jstor.com/stable/41680007

[21] MacKay NJ. When Lanchester met Richardson, the outcome was stalemate: A parable for mathematical models of insurgency. Journal of the Operational Research Society. 2015;**66**(2):191-201. Available from: http://www.jstor.com/stable/24505286

[22] Kress M. Lanchester models for irregular warfare. Mathematics. 2020;**8**:737. DOI: 10.3390/math8050737

[23] MacKay N, Price C. Safety in numbers: Ideas of concentration in Royal air Force Fighter Defence from Lanchester to the Battle of Britain. History. 2011;**96**(3):304-325. Available from: http://www.jstor.com/stable/24429278

[24] Adams ES, Mesterton-Gibbons M. Lanchester's attrition models and fights among social animals. Behavioral Ecology. 2003;**14**(5):719-723. DOI: 10.1093/beheco/arg061

[25] Plowes NJR, Adams ES. An empirical test of Lanchester's square law: Mortality during battles of the fire ant *Solenopsis invicta*. Proceedings: Biological Sciences. 2005, 2005;**272**(1574):1809-1814. Available from: http://www.jstor.com/stable/30047764

[26] Möbius strip. Wikipedia. Available from: https://tinyurl.com/3jrs2xvt

[27] Möbius mayhem 2123 - 'Vernichtungsschlacht' in the vacuum of space. Wargames Vault. 2023. Available from: https://tinyurl.com/mu7bp4en The game is free

[28] Flanagan M, Lipscombe TC, Northey A, Robinson IM. Chance all—A simple 3D6 dice stopping game to explore probability and risk vs reward. London, UK, Intechopen: Game Theory —From Idea to Practice; 2022. Available from: https://www.intechopen.com/chapters/83376

[29] Armstrong MJ. The salvo combat model with a sequential exchange of fire. Journal of the Operational Research Society. 2014;**65**(10):1593-1601. Available from: http://www.jstor.com/stable/24505020

[30] Online gifs of games in Section 3. Available from: https://iactaaleaest.wordpress.com/2003/01/17/lanchesters-fighting-strength-as-a-battle-outcome-predictor-applied-to-a-simple-fire-and-manoeuvre-wargame-supporting-gifs/

[31] Flanagan M, Northey A, Robinson IM. Exploring tactical choices and game design outcomes in a simple wargame 'take that hill' by a systematic approach using experimental design. International Journal of Serious Games. 2020;7(4):27-50. DOI: 10.17083/ijsg.v7i4.372

[32] Take that Hill! UK Fight Club's First Manual Wargame Primer. Available from: https://www.ukfightclub.co.uk/take-that-hill

[33] Plackett RL, Burman JP. The design of optimum multifactorial experiments. Biometrika. 1946;**33**(4):305-325. DOI: 10.1093/biomet/33.4.305

[34] Christy DE. A Lanchester Based Model for Analyzing Infantry Fire and Maneuver Tactics [M.S. in Operations Research]. Naval Postgraduate School; 1969. Available from: http://hdl.handle.net/10945/12536

[35] Taylor JG. Lanchester-Type Models of Warfare. Volume I. Naval Postgraduate School. 1980. ADA090842. Available from: https://apps.dtic.mil/sti/pdfs/ADA090842.pdf

[36] von Clausewitz C. Book 4, Chapter 4. In: On War. 1832. Available from: https://tinyurl.com/2p9hc78z

[37] Battle of annihilation. Wikipedia. Available from: https://tinyurl.com/yjw57u92

[38] Dolanský L. Present state of the Lanchester theory of combat. Operations Research. 1964;**12**(2):344-358 Available from: http://www.jstor.com/stable/167934

[39] Anderton CH. Toward a mathematical theory of the offensive/defensive balance. International Studies Quarterly. 1992;**36**(1):75-99. Available from: http://www.jstor.com/stable/2600917

[40] Kuikka V. A combat equation derived from stochastic modeling of attrition data. Military Operations Research. 2015;**20**(3):49-69. Available from: https://www.jstor.org/stable/10.2307/24838621

[41] Taylor JG. Optimal commitment of forces in some Lanchester-type combat models. Operations Research. 1979;**27**(1):96-114. Available from: http://www.jstor.com/stable/170246

[42] Kingman JFC. Stochastic aspects of Lanchester's theory of warfare. Journal of Applied Probability. 2002;**39**(3):455-465. Available from: http://www.jstor.com/stable/3216070

[43] Brooks FC. The stochastic properties of large Battle models. Operations Research. 1965;**13**(1):1-17. Available from: http://www.jstor.com/stable/167950

[44] Brown RH. Theory of combat: The probability of winning. Operations Research. 1963;**11**(3):418-425. Available from: http://www.jstor.com/stable/168029

[45] Brackney H. The dynamics of military combat. Operations Research. 1959;**7**(1):30-44. Available from: https://www.jstor.org/stable/167591

Chapter 2

A New Monte Carlo Approach for the Solution of Semi-Linear Neumann Boundary Value Problem

Abdujabar Rasulov

Abstract

In this paper we will consider the Neumann boundary-value problem for the Laplace operator with a polynomial nonlinearity on the right-hand side. We utilize Greens formula and an elliptic mean-value theorem. This allows us to derive a probabilistic representation as special integral equation which equates the value of the function u(x) at the centre-point x, with its integral over the domain D, and on boundary of the domain G. According to this probabilistic representation, a Markov branching process is constructed and an unbiased estimator of the solution of the nonlinear problem is formed by taking the mathematical expectation over these branching processes. The unbiased estimator which we derive has a finite variance. The complexity of the proposed algorithms is investigated, and the ways decreasing of the computational work are proposed. Some particular cases are considered in detail. The effectiveness of two estimators "by absorption" and "by collision" is studied in the nonlinear case.

Keywords: Neumann boudary value problem, Monte Carlo method, Markov branching process, martingale, unbiased estimator

1. Introduction

In this work, we will propose new probabilostic algorithms for the solution of the Laplace operator with a polynomial nonlinearity on the right-hand side. The complexity of the proposed algorithms is investigated, and the ways decreasing of the computational work are proposed. For the solution of the semi-linear Neumann problem, a new and effective Monte Carlo algorithm is derived. Some particular cases are considered in detail. The effectiveness of two estimators "by absorption" and "by collision" is studied in the nonlinear case.

2. Solutions of semi-linear Neumann problems

Assume D is a bounded convex domain in $R^n, n \geq 3$ with a regular smooth boundary Γ, and $x = (x_1, \dots, x_n) \in D$. Consider the following boundary value problem:

$$
\begin{aligned}
&Lu(x) \equiv -\Delta u(x) + g(x)u(x) = f(x, u(x)), \\
&\left.\frac{\partial u(x)}{\partial n_x}\right|_{\Gamma} = 0.
\end{aligned} \tag{1}
$$

where n_x is the external normal on the boundary Γ at the point x, and the function $g(x) > 0$ and $g(x) \in C(\overline{D})$. We shall determine the class of functions

$$
v = \{\omega(x) : B^* M(x, 0) + \psi(xt) \geq 0, \text{ in } D\}; \tag{2}
$$

Here $M(x,y)$ is the Levy function for operator L (see: [1]), $\psi(x)$ is a continuous function, which has continuous derivatives of second order in $D \cup \Gamma$, and B is a positive constant. It is known, that this class of functions is not empty.

Suppose that the function $f(x, \omega(x))$ satisfies the following three conditions:

a. $\partial f(x, \omega)/\partial\omega \geq 0$ is an increasing function at $\omega \geq 0$;

b. There exists such k, that $f(x, \omega) = o\left(r^{k-2}\right)$ and $\frac{\partial f(x, \omega)}{\partial \omega} = o\left(r^{k-1}\right)$, here r is the distance from the origin to the point x;

c. The limit $\lim\limits_{x \to 0}\left[\omega(x) h^{-1}(x,y)\right] = B$ holds, where $y = \left(y_1, \dots, y_n\right) \in D$ and

$$
h(x,y) = \frac{1}{(-1)^{n/2}(n-2)\,\omega_n}\left[(-1)^n \sum_{i=1}^{n} x_i - y_i)^2\right]^{(2-n)/n}, \tag{3}
$$

and ω_n is a surface element of the unit sphere.

The next assertion is then true.

Lemma 1 *The unique positive solution of (1) exists within the class of functions, v, if the above conditions a), b), and c) are satisfied.*

This assertion follows from the results of [2], therefore the proof is omitted. Thus, there exists a unique function $\omega(x) = u(x) \geq 0$ (from lemma 1) that satisfies conditions (a), (b), and (c) and which is the solution of problem (1).

Let us construct a computational scheme for the solution of problem (1) using proposed algorithms from the work [3]. Suppose Γ is a sufficiently smooth boundary and the function $v(x) \in C^2(D) \cap C^1(D)$. Then, we can use Green's identity to obtain

$$
\int_D (v(y)Lu(y) - u(y)Lv(y))dy = \int_\Gamma u(y)\frac{\partial v(y)}{\partial n_y} - v(y)\frac{\partial u(y)}{\partial n_y} dy\Gamma \tag{4}
$$

Let us $c^2 > \max\limits_{x \in \overline{D}} g(x)$. then using formula (4) we can derive the integral equation for $u(x)$

$$
\begin{aligned}
u(x) = {} & \frac{1}{\omega_n}\left(\int_D \frac{\exp(-c\rho)}{\rho^{n-2}} f(y), u(y)\right) dy + \\
& + \int_D \frac{A_2(\rho) - \frac{g(y)}{n-2}\rho\exp(-c\rho)}{\rho^{n-1}} u(y)dy + \int_\Gamma A_1(\rho)\frac{\cos\varphi_{xy}}{\rho^{n-1}} d_y\Gamma.
\end{aligned} \tag{5}
$$

Here φ_{xy} is the angle between the normal n_y with the vector $y - x$, $\rho = |x - y|$,

$$A_1(\rho) = \left(1 + \frac{c\rho}{n-2}\right)\exp(-c\rho), \quad A_2(\rho) = \frac{c^2\rho - c(n-3)}{n-2}\exp(-c\rho). \tag{6}$$

It is easy to show that $(A_1(\rho))' = -A_2(\rho), \ \ A_1(0) = 1$. Thus,

$$\int_0^\rho A_2(\rho)d\rho = 1 - A_1(\rho) \tag{7}$$

Furthermore, if $x \in \Gamma$, then ω_n should be changed into $\omega_n/2$.

The linear case, when $f(x, u(x)) = f(x)$ was considered by A.Sipin. In his work [4] sequences of unbiased estimators were developed and the question of simulating the required special densities was discussed. Furthermore, using the results of [4], we will present an algorithm to construct of unbiased estimators for the solution of problem (1).

For $u(x)$, after transforming Eq. (1) as discussed below, we obtain the following integral equation

$$\begin{aligned} u(x) = & \int_D (1 - \pi(x,y))\Pi(x,y)u(y)d\mu(y) + \\ & + \frac{1+\chi_\Gamma(x)}{\omega_n}\int_D f(y,u(y))\frac{\exp(-c\rho)}{\rho^{n-2}}d\lambda(y). \end{aligned} \tag{8}$$

Here

$$\chi_\Gamma(x) = \begin{cases} 0, & x \notin \Gamma, \\ 1, & x \in \Gamma. \end{cases} \quad \pi(x,y) = \begin{cases} 0, & y \in \Gamma \\ \dfrac{g(y)\rho\exp(-c\rho)}{(n-2)A_2(\rho)}, & y \in D \end{cases}$$
$$\Pi(x,y) = \frac{1+\chi_\Gamma(x)}{\omega_n}\begin{cases} \left(A_1(\rho)\cos\varphi_{xy}\right)/\rho^{n-1}, & y \in \Gamma, \\ A_2(\rho)/\rho^{n-2}, & y \in D \end{cases} \tag{9}$$

The measure, μ, is determined by the σ-algebra of Borel subsets $\overline{D}$ with the equality $\mu(E) = \lambda(E) + S(E \cap \Gamma)$, where Lebesque λ-measure is in R^n, S is a surface element. It is easy to see that $\Pi(x,y) \geq 0$. Because the domain D is convex, equality (7) implies that $\Pi(x,y)$ is a transition probability density. Assume $q(x) \not\equiv 0, \ c^2 > \max_{x \in \overline{D}} g(x)$. Then, $0 \leq \pi(x,y) \leq 1$. Let now consider

$$f(x,u(x)) = \sum_{i=0}^{m} b_i(x)u^i(x), \ \ x \in \overline{D}, \tag{10}$$

where $b_i(x) \geq 0$. We choose $\alpha_i, \quad i = 0, \ldots, m+1$, to satisfy the condition:

$$\sum_{i=0}^{m+1} \alpha_i = 1, \ \ \sum_{i=1}^{m} i\alpha_i + \alpha_{m+1} \leq 1 \tag{11}$$

The iterative method for the solution of Eq. (8) will converge only with small coefficients, $b_i(x)$. Therefore, we assume that parameters α_i satisfy the following condition for each $i, \quad 0 \le i \le m+1$:

$$\|b_i(x)\|_{C(\overline{D})} \le 2\alpha_i(n-1)/d^2, \tag{12}$$

here d is diameter of the domain D, and $\alpha_i \not\equiv 0, \quad i = 0, \ldots, m$. Using (10) and (11) we can transform Eq. (8), as follows

$$\begin{aligned} u(x) = {} & \alpha_{m+1} \int_{\overline{D}} k_{m+1}(x,y)\Pi(x,y)u(y)d\mu(y) + \\ & + \frac{1+\chi_\Gamma(x)}{\omega_n} \sum_{i=1}^{m} \alpha_i \int_D \frac{K_i(x,y)}{\rho^{n-2}} u^i(y)d\lambda(y) + \\ & + \frac{1+\chi_\Gamma(x)}{\omega_n} \alpha_0 \int_D \frac{K_0(x,y)}{\rho^{n-2}} d\lambda(x) \end{aligned} \tag{13}$$

where

$$\begin{aligned} K_{m+1}(x,y) = (1-\pi(x,y))/\alpha_{m+1}, \quad k_0(x,y) = b_0(y)\exp(-c\rho)/\alpha_0, \\ K_i(x,y) = b_i(y)\exp(-c\rho)/\alpha_i, \quad i = \overline{1, \ldots, m} \end{aligned} \tag{14}$$

We can use a method based on branching random process to Eq. (13), by applying the method for integral equations with polynomial nonlinearities.

We define a branching Markov process in $\overline{D}$, and a sequence of unbiased estimators on the trajectory of the process as follows. Assume at the initial time and at x there is only one particle. With probability α_0 the particle moves to the point $y \in D$, where $y = x + \rho\overline{\omega}$ where $\overline{\omega}$ is an isotropic vector in R^n (if $x \in D$) and ρ is distributed uniformly in $[0; d]$ with density $2\rho/d^2$. If $x \in \Gamma$, then $\overline{\omega}$ is an isotropic vector in the half-space G, in which D lies. At the point y, the particle is absorbed. In this case, the estimator is

$$\xi_1(x) = \frac{d^2}{2(n-2)} \exp(-c\rho)\tilde{b}_0(x+\rho\overline{\omega})/\alpha_0, \tag{15}$$

where

$$\tilde{b}_0(x) = \begin{cases} b_0(x), & x \in \overline{D} \\ 0, & x \in \overline{D} \end{cases} \tag{16}$$

With probability $\alpha_i, \quad i = 1, \ldots, m$, the particle at point x moves to $y = (x + \rho\overline{\omega})$ (here ρ is as above) and at y we generate i particles.

In this case, the estimator will be

$$\xi_1(y) = \left(\frac{d^2}{2(n-2)} \exp\left(-c\rho\tilde{b}(x+\rho\overline{\omega})/alpha_i\right) u^i(x+\rho\overline{\omega}), \tag{17}$$

here

$$\tilde{b}_i(x) = \begin{cases} b_i(x), & x \in \overline{D} \\ 0, & x \text{-} \in \overline{D} \end{cases} \tag{18}$$

With probability α_{m+1}, the particle moves to the point y, which is distributed with density $\Pi(x,y)$, and generates one particle. The estimator is thus

$$\xi_1(x) = ((1 - \pi(x,y))/\alpha_{m+1})u(y_1). \tag{19}$$

The algorithm for simulating a random point with density functions $\Pi(x,y)$ is the following [3]:

1. Let $n \geq 0$ and x_n be known.

2. If $x_n \in G$, then Π_{x_n} denote a half-space in which $D \in \Pi_{x_n}$, and is defined by the tangent plane to D at the point x_n.

3. Let ω is an isotropic vector in Pi_{x_n}, if $x_n \in G$ and an isotropic vector in R^m, if $x_n \in D$

4. $\rho = \rho(x_n, \omega) = sup\{t > 0 : x_n + t\omega\}$ is the length of a ray, that emitting from the point x_n and lies in D pointing in the ω direction. $x_n + \rho\omega$ is always a point on the boundary.

5. The Markov chain jumps from x_n to x_{n+1}, where $x_{n+1} = \xi = x_n + \rho\omega$ with probability $A_1(\rho)$, and with probability $1 - A_1(\rho)$ ξ is distributed on the segment $[x_n, x_n + \rho\varpi]$ with the probability density $p_1(r) = A_2(\rho)/(1 - A_1(\rho))$.

The simulation method, which was defined above gives a vector ξ, which is distributed *via* $\Pi(x,y)$.

It is not difficult (see Ref. [3]), that $\xi_t(x)$ is a martingale relative to the σ-algebra $\{F_t\}_{t=1}^{\infty}$, where F_t was generated by the process up to time t. Because $\xi_t(x) \geq 0$ (from lemma 1), $\xi_z(x)$ is a supermartingale, then (see Ref. [5], p. 112) with probability one if tends to the integrable random variable $\xi_\infty(x)$. To show that $Ex|\xi_\infty(x)| < +\infty$ and $xi_n(x) = Ex\{\xi_\infty(x)/F_n\}$, we must show that $\xi_t(x)$ is uniformly integrable. We will show that our estimator is bounded and that $\xi_\infty(x)$ is an unbiased estimator for $u(x)$. It is known [5] that if $Ex \;\; xi_t^2(x) < +\infty$, then $\xi_t(x)$ is uniformly integrable.

Theorem 1 *If condition* (12) *is satisfied and*

$$\alpha_{m+1} \geq \max_{(x,y) \in D} \left\{ \left(c^2 - g(y)\rho + c(n-3)\right) / \left(\left(c^2 p^2 + c(n-3)\right)\right) \right\} \tag{20}$$

then $\xi_t(x)$ is a supermartingale and tends to integrable random variable $\xi_\infty(x)$ almost surely.

Proof: From the definition of our estimator, it is easy to see that if we choose α_{m+1} to satisfy the inequality below

$$\alpha_{m+1} \geq \max_{x,y \in D} \left\{ \frac{(c^2 - g(y))\rho + c(n-3)}{c^2\rho + c(n-3)} \right\} \tag{21}$$

Then, $E_x\xi_t^2(x) < +\infty$ from condition (12). From (21) one can see that the inequality $0 < \alpha_{m+1} \leq 1$ holds. The parameter c we could choose, so that the inequality (21) is less than one because $g(y) > 0$. The rest of the α_i can be chosen differently from zero. □.

Thus, $\xi_\infty(x)$ is an unbiased estimator for $u(x)$ at $x \in D$. Here, the estimator $\xi_\infty(x)$ is more amenable to practical computation. It differs from the estimator used for the nonlinear Dirichlet problem, and it is not necessary to create an ε -neighborhood of the boundary.

3. The case when $n = 2$ and $g(x) = c^2$

Let D be a bounded domain in R^3, with a regular smooth boundary Γ. Consider the following nonlinear problem

$$\begin{aligned} &-\Delta u(x) + c^2 u(x) = b(x)u^2 + f(x) \\ &\left.\frac{\partial u(x)}{\partial n_x}\right|_\Gamma = 0 \end{aligned} \tag{22}$$

Since $A_2(\rho) = c^2\rho\exp(-c\rho)$ the volume integral in (5) vanishes. In this particular case, we obtain the following integral equation

$$\begin{aligned} u(x) = &\frac{1+\chi_\Gamma(x)}{4\pi}\int_D (\exp(-c\rho)/\rho)b(y)u^2(y)dy+ \\ &+\frac{1+\chi_\Gamma(x)}{4\pi}\int_D (\exp(-c\rho)/\rho)f(y)dy+ \\ &+\frac{1+\chi_\Gamma(x)}{4\pi}\int_\Gamma (1+c\rho)\exp(-c\rho)\left(\cos\varphi_{xy}\right)/\rho^2)u(y)dy\Gamma \end{aligned} \tag{23}$$

We denote.

$\pi(x,y) = 1 - \exp(-c\rho)(1+c\rho)$, and $p_1(x,y) = \frac{1+X_\Gamma(x)}{4\pi\rho^2}\cos\varphi_{xy}$ and transform the above integral equation to get

$$\begin{aligned} u(x) = &\frac{1+X_\Gamma(x)}{2}d^2\int_D (2\rho\exp(-c\rho)/4\pi d^2\rho^2)b(y)u^2(y)dy \\ &+\frac{1+\chi_\Gamma(x)}{2}d^2\int_D (2\rho\exp(-c\rho)/4\pi d^2\rho^2)f(y)dy+ \\ &+\int_\Gamma (1-\pi(x,y))\quad p_1(x,y)u(y)d_y\Gamma \end{aligned} \tag{24}$$

From Eq. (24) it is obvious how to construct the branching Markov process and the sequence of unbiased estimators analogous to those determined above [6, 7]. Here the difference is that we do not introduce artificial parameters α_i. Usually, when the parameters α_i are introduced, the estimation will change, and sometimes, if we are not careful, the variance of the estimators may be unbounded.

Assume $b(x) \geq 0, f(x) \geq 0$ are such that

$$\max\left\{\|b(x)\|_{C(\overline{D})}, \|f(x)\|_{C(\overline{D})}\right\} \leq \min_{x,y \in D}\left\{\sqrt{\pi(x,y)/d^2}\right\} \quad (25)$$

Here d is the diameter of the domain D.

Now, we will determine the branching Markov process in $\overline{D}$ and the corresponding sequence of unbiased estimators. Assume that at x there is only one particle. Furthermore, the point y is distributed with density $p_1(x,y)\, y \in \Gamma$. With probability $1-\pi(x,y)$, we generate a particle at y. In this case, the estimator is $\xi_1(x) \equiv 1 \cdot u(y)$. With probability $\pi(x,y)/2$, the particle moves from x to the point $y_1 = x + \rho\overline{\omega}$ and is absorbed. Here, $\overline{\omega}$ an isotropic vector in R^3, and ρ is distributed in $[0,d]$ with density $2\rho/d^2$. In this case, the estimator will be

$$\xi_1(x) = \left(d^2 \exp(-c\rho)\pi(x,y)\tilde{f}(y_1)\right), \quad (26)$$

where

$$\tilde{f}(y_1) = \begin{cases} f(y_1), & y \in \overline{D}, \\ 0, & y \notin \overline{D}. \end{cases} \quad (27)$$

Two particles are created at y_1 with probability $\pi(x,y)/2$. In this case, the estimator is

$$\xi_1(x) = \left(d^2 \exp(-c\rho)\overline{b}(y_1)/\pi(x,y)\right)u^2(y_1), \quad (28)$$

where

$$\tilde{b}(y_1) = \begin{cases} b(y_1), & y_1 \in \overline{D}, \\ 0, & y_1 \notin \overline{D}. \end{cases} \quad (29)$$

New particles behave in a manner similar to their parents. $\xi_t(x) = \xi_{t-1}(x)$ if the process terminates up to time t, and $\xi_t(x)$ is obtained from $\xi_{t-1}(x)$ by the replacing $u(y)$ in the estimators by $\xi_1(x)$. Obviously, $\xi_t(x)$ is a martingale relative to the family of $\sigma-$ algebras, $\{F_t\}_{t=1}^{\infty}$, where F_t is generated by the process up to time t.

From the conditions (25), it is easy to see that $E_x\xi_t^2(x) < +\infty$, so that $\xi_t(x)$ is a uniformly integrable martingale. Therefore, $\xi_t(x)$ tends to $\xi_\infty(x)$ almost surely, and $E_x|\xi_\infty(x)| < +\infty$, and $\xi_n(x) = E_x\{\xi_\infty(x)/F_n\}$. Thus, $\xi_\infty(x)$ is the unbiased estimator for solution to problem (22) at the point $x \in D$.

Note that m_1 is the average number of particles generated over one time stop. In our case, $m_1 = 1$ and so the constructed branching Markov process is critical in the sense of neutronics.

The next assertion is true.

Theorem 2 *With probability one, the constructed branching Markov process terminates.*

The proof is analogous to that of theorem [8] and is thus omitted.

We should note that the above method for constructing unbiased estimators may be used for solving mixed problems when parts of the boundary are given a Dirichlet condition and other parts of Neumann condition. It is only important, that pieces of the boundary, where the normal derivative exists, should be convex surfaces. In this case, we use the estimators with a combination of the above proposed algorithms with the process "walk on spheres with branching" (see Ref. [9]).

It is not difficult to see, that these estimators use namely absorption. Depending on the complexity of algorithms, one may use the other types of estimators (e.g., collisions estimators) (see Ref. [10]).

4. Comparison of the complexity of the estimators absorption and collision in the quasi-linear case

The comparison of the complexity absorption and collision estimators in the quasi-linear case is a difficult problem. However, in some particular cases, one can analyze the complexity in this case.

To optimize the calculation procedure we can sometimes choose different Markov chains and estimators. In this case, the choice of the estimator is most important and could increase the efficiency of the algorithm. It is often useful to compare two estimators and chose one of them base on the particular problem being solved. We note that the computational cost may depend on the type of equation and the type of Markov process. It may turn out that one estimator is best in one situation while the other maybe best for a difficult equation.

For quasi-linear equations we will give some recommendations in choosing estimators that minimize the complexity of a calculation (see Ref. [11]).

Let consider the following nonlinear integral equation

$$\varphi(x) = f(x) + \\ + \sum_{m=1}^{n} \int_D \cdots \int_D K_m(x, y_1, \ldots, y_m) \prod_{i=1}^{m} \varphi(y_1) dy_1 \ldots dy_i \tag{30}$$

where $D \subset R^n$, $K_i(x, y_1, \ldots, y_m)$, and $f(x)$ are given functions in n times $D \times \cdots \times D$, and $\varphi(x)$ is the unknown function. These problems, which are connected with the construction of unbiased estimators for the functional (h, φ), where $h(x)$ is a given function and was discussed in Refs [10, 12]. We assume that $K_i(x, y_1, \ldots, y_i)$, and $f(x)$ satisfy the condition in [10, 12].

It's known that the solution of nonlinear integral equations by Monte Carlo is connected with branching Markov processes, determined by the initial distribution density $P_0(x) \geq 0$, and the transition density $P_i^1(x, y_1, \ldots, y_i)$. Here

$$\int_0 P_0(x) dx = 1, \tag{31}$$

$$\sum_{i=1}^{n} \int_D \cdots \int_D P_i^1(x, y_1, \ldots, y_i) dy_i \ldots dy_i = 1 - g^1(x), \tag{32}$$

where $g^1(x) \in [0, 1)$ is the probability of terminating the branching Markov process trajectory. Realizations of this process are connected with "trees," on which were constructed different unbiased estimators.

We consider only estimators using absorption or collision and will give recommendations for their use based on complexity.

As is known [13], the integral equation

$$a(x) = \sum_{m=1}^{n} \int_D \dots \int_D P_m^1(x, y_1, \dots, y_m) \prod_{i=1}^{m} a(y_i) dy_1, \dots dy_i + g^1(x) \quad (33)$$

has a unique positive solution for all $x \in D$ which is determined through a Neumann series. With probability one, the number of generations in the branching Markov process is bounded.

If the branching Markov process in the branches splits into m- new particles with given probabilities α_m, and the probability of absorption is $g^1(x) = \alpha_0$, then probabilities α_m satisfy the conditions.

$$\sum_{i=1}^{n} \int_D \dots \int_D \alpha_i P_i(x, y_1, \dots, y_i) dy_1 \dots dy_i = 1 - \alpha_0, \quad (34)$$

where the functions $P_1(x, y_1, \dots, y_i)$ satisfy

$$\int_D \dots \int_D P_i(x, y_1, \dots, y_i) dy_1 \dots dy_i = 1. \quad (35)$$

It is not difficult to see that (see sec.1) that the function $M(x)$, which is the mean number of particles in the branching process, satisfies the linear integral equation

$$M(x) = \alpha_0 + \sum_{i=1}^{n} \alpha_i \int_D \dots \int_D P_i(x, y_1, \dots, y_i) \sum_{j=1}^{i} M(y_j) dy_1 dy_i. \quad (36)$$

Here $M(x)$ the iterative solution to this equation with the initial value α_0. It's known (see Section 1) that the constructed branching Markov process will be terminated in condition $b = \sum_{m=1}^{n} m\alpha_m \leq 1$, this is why we choose α_m so that $b \leq 1$, when b is the average number of particles generated in one step.

We solve the integral Eq. (30) at the point $x \in D$. When the function h is a Dirac δ-function, that is $\int h(x-y)\varphi(x)dx = \varphi(y)$. In this case, at the initial time, we have one particle at the point x almost, surely. The branching process is constructed on a tree of branches then we will construct unbiased estimators using absorption, $\xi_n(x)$, and collision, $\xi_c(x)$. To compare the complexity of these we compute on this tree.

Obviously, the number of arithmetic operations depends on the average number of particles in the tree. Furthermore, we will denote D_n as the variance of the absorption estimator, D_c and the variance of the collision estimator. The complexity of the absorption estimator is $T_n = t' \cdot D_n$ and the complexity collision estimator $D_c = t'' \cdot D_c$. Here t' and t'' as the average CPU time for calculation estimators, which will be needed for calculation T_n and T_c.

First, we will investigate the variance of both estimators.

Let F^+ and F^- be a set of positive and negative bounded functions. We shall consider the following integral equation

$$\psi(x) = \frac{f^2(x)}{\alpha_0} + \\ + \sum_{m=1}^{n} \int_D \cdots \int_D \frac{K_m^2(x, y_1, \dots, y_m)}{\alpha_m p_m(x, y_1, \dots, y_m)} \prod_{i=1}^{m} \psi(y_i) dy_1, \dots, dy_i \tag{37}$$

$$\psi_1(x) = f(x)(2\varphi(x) - f(x)) + \\ + \sum_{m=1}^{n} \int_D \cdots \int_D \frac{K_m^2(x, y_1, \dots, y_m)}{\alpha_m p_m(x, y_1, \dots, y_m)} \prod_{i=1}^{m} \psi_1(y_i) dy_1, \dots, dy_i \tag{38}$$

We suppose that the minimal solutions of Eqs. (37), (38) exist and that Neumann series converges for (37) with the initial function $\psi^{(0)}(x) = \frac{f^2(x)}{\alpha_0}$ and for (38) with the initial function $\psi^{(0)}(x) = f(x)(2\varphi(x) - f(x))$. Then the expression for the variances will be

$$D_n = (h^2/p_0, \psi) - (h, \varphi)^2, \quad D_c = (h^2/p_0, \psi_1) - (h, \varphi)^2. \tag{39}$$

Let the function $G(x) \in F^+$, where $G(x) = f(x)(2\varphi(x) - f(x))$. Moreover, if $f(x) \geq 0$ then $G(x) \in F^+$, in this case, the following hold:

Theorem 3 *Let $f(x) \in F^+$. If the probability of terminating the trajectory of our branching processes is*

$$\alpha_0 \in [0, \min G(x), x \in D], \tag{40}$$

then $D_c \leq D_n$.

Proof: Since the values of D_n and D_c depend on solutions of Eqs. (37), (38), we will investigate the minimal solutions of these equations. The minimal solutions of these equations continuously depend on right-hand sides of the equations. Consequently, if $\alpha_0 < \min(f(x)(2\varphi(x) - f(x)))$, then $\psi_1(x) \leq \psi$. From here, $D_c \leq D_n$ □.

Consequence: If it follows that $\alpha_0 \notin [0, \min G(x), x \in D]$, then $D_n \leq D_c$.

Theorem 4 *Suppose that $f(x) \in F^-$. If the probability of terminating a trajectory of our branching processes is*

$$\alpha_0 \in [0, \min G(x), x \in D], \tag{41}$$

then $D_n \leq Dc$.

The proof is analogous to the above.

Thus, using a priori information about the variance of estimators (solutions of the Eqs. (37), (38)), one can give recommendations on choosing optimal estimators.

Author details

Abdujabar Rasulov
Department System Anasysis and Mathematical Modeling, University of World Economy and Diplomacy, Tashkent, Uzbekistan

*Address all correspondence to: asrasulov@gmail.com

References

[1] Miranda C. Partial Differential Equations of Elliptic Type. Berlin-Heidelberg-New York: Springer Verlag; 1970. p. 372

[2] Anishenko RN. Theorems of existence and method of sequential approximations for some nonlinear problems "Russian mathematics", (Izvestiya Vysshikh Uchebnykh Zavedenii-in Russian). Mathematica. 1971;**5**:65-64

[3] Ermakov SM, Nekrutkin VV, Sipin AS. Random Processes for Classical Equations of Mathematical Physics. Dordrecht, Netherlands: Springer; 2013. p. 302

[4] Sipin AS. About the solution of Neyman problem by Monte-Carlo method. In: Metodi Monte-Karlo v vichislitelnoy matematike i matematicheskoy fizike. Novosibirsk: Computing center of RAN publisher; 1976. pp. 120-136

[5] Meyer PA. Probability and Potentials. Waldham, Massachusetts, Toronto, London: Blaisdell Publishing Co.; 1966. p. 266

[6] Rasulov AS. The construction of unbiased estimators for solving one nonlinear problem of Neyman. In: Algoritmi i chislennie metodi resheniya prikladnoy mehaniki. Tashkent: Trudi Tash GU; 1986. pp. 64-68

[7] Rasulov AS, Raimova GM. Monte Carlo solution of the Neumann problem for the nonlinear Helmholtz equation. Mathematics and Computers in Simulation. 2015;**117**:1-9. DOI: 10.1016/j.matcom.2015.05.002

[8] Rasulov AS. Monte Carlo algorithms for the solution quasi-linear Dirichlet boundary value problems of elliptical type. Journal of Mathematics ans Statistics (to appiar-submitted). 2023; **11**(3):592-597

[9] Puskarev LI, Rasulov AS. Solution of some boundary value problems with nonlinearity in boundary conditions by random walks on spheres. In: Problems of Computational and Applied Mathematics. Vol. 61. Tashkent: TashGU publisher; 1980. pp. 47-52. (in Russian)

[10] Sizova AF. About the Variance of Estimator by Collisions for Solving Nonlinear Problems by Monte-Carlo Method. Vol. 1. Saint Petersburg: Vestnik LGU; 1976. pp. 152-155

[11] Rasulov AS. The efficiency of two models of Monte-Carlo method for solving nonlinear problems. In: Prikladnaya matematika i mehanika. Tashkent: Trudi Tash GU; 1983. pp. 60-64

[12] Ermakov SM. The analogue of Neyman-Ulam scheme in nonlinear case. Jurnal vichislitelnoy matematiki i matematicheskoy fiziki. 1973;**13**(3): 564-573

[13] Ermakov SM. Monte-Carlo method for iteration of nonlinear operators. DAN SSSP. 1972;**204**(2):271-274

Chapter 3

The Effect of Anode, Hot and Cold Cathode Structures Used in Miniature X-Ray Tubes on X-Ray Output Using Monte Carlo Simulation

Barat Barati, Masumeh Kaydani, Fariba Farhadi Birgani, Mansour Zabihzadeh, Mohammad Javad Tahmasebi and Nahid Chegeni

Abstract

This study aimed to investigate the use of carbon layers (cold cathode) as an alternative to tungsten metal (hot cathode) for electron production in miniature X-ray tubes in the field of electronic brachytherapy. The structural impact of these tubes on X-ray production was also analyzed. Using a Monte Carlo code, hemispherical-conical, hemispherical, and truncated-conical miniature X-ray tubes were simulated. Both cold and hot cathodes were utilized as electron sources to calculate the X-ray spectrum and dose in two directions. The results showed that the ratios of axial and transverse doses of spherical to cylindrical cold cathodes were respectively 5.4, 4.8, and 4.6, and 2.4, 2.2, and 1.9 in the hemispherical-conical, hemispherical, and truncated-conical miniature sources. In conclusion, the structure of the hemispherical-conical miniature X-ray tubes demonstrated better performance compared to the hemispherical and truncated-conical tubes. The different shapes of the cathode and anode in miniature X-ray tubes had a significant impact on the generation of output X-rays.

Keywords: hot and cold cathode, Monte Carlo, miniature X-ray tube, electronic brachytherapy, anode

1. Introduction

Electronic brachytherapy [1] is a new form of electrotherapy using X-rays to deliver a dose of radiation to target tissues at a distance of a few centimeters through the intracavitary, intraluminal, or interstitial approaches [2]. In X-ray tubes with a hot tungsten cathode create a high potential difference between the cathode and the anode. The electrons produced in the hot tungsten cathode collide with the anode, leading to generation of an X-ray. The cathode current density, energy consumption, and response time play important roles in the production of X-rays. One of the major hot cathodes is the tungsten filament [3]. The current density equation of the tungsten filament is given in the Appendix 1 [4]. The hot tungsten cathode can work even in non-ultra-high vacuum environments containing a large number of gas molecules. However, hot cathodes are exposed to chemical reactions that generate tungsten oxides, which become thinner over a long period by sublimation of oxides [5]. Therefore, X-ray production with this method faces limitations such as gradual response time, high energy consumption, and time limit.

In the mid-1950s, replacing cold cathodes with hot cathodes was suggested to overcome the disadvantages of hot cathodes [5]. Accordingly, the X-ray production using the field emission method was first reported in 1956 [6]. The necessary equations and explanations in this field are given in Appendix 2 [7]. The carbon layer can be used as a cold source to generate the flow of free electrons so that the electrons can be quickly removed from the surface. As a result, no oxidation reaction or deformation is required in the appearance of the carbon layer. Equations A1 and A2 shows that carbon and tungsten layers both have the same work function (Φ = 4.5ev). Tungsten is currently used as an electron generator in conventional X-ray tubes and miniatures, despite its limitations. Since the work functions of tungsten and the carbon layer are the same, in this study, the X-ray output of a cold carbon cathode was evaluated in comparison with the X-ray output of a hot tungsten cathode. Next, the effect of the geometry of different cathodes and anodes on this issue was discussed.

2. Methods and materials

2.1 Simulation of miniature X-ray tubes and the MCNP code

In this study, hemispherical-conical, hemispherical, and truncated-conical miniature X-ray tubes [6, 8] simulated with the Monte Carlo code MCNPX (2.6.0) were used (**Figure 1**). In these simulations, the densities of tungsten, beryllium, and water were 19.3, 1.84, and 0.998 g/cm^3 according to the Reference [7], and the density of the carbon layer was 1.35 gr/cm^3 according to the Reference [8, 9]. The F6 tally was applied to calculate random heat as an absorbed dose estimator. We simulated the transport of photons and electrons (Mode: P E) with the cut-off energy of 1 Kev 1). According to **Figure 2**, each miniature X-ray tube included two main parts called the electron production source (cathode) and anode (the electron collision site). Thus, the hemispherical (r = 0.8 mm radius), cylindrical (height = 0.8 mm and r = 0.8 mm), and disk-shaped (r = 0.8 mm and thickness = 0.1 mm) cathode structures were used in all stages of the work. The name of the tube indicates the anode structure. In all phases of the study, miniature tubes were placed in the spherical water phantom.

2.2 Calculation of the dose using cold carbon cathode (cylindrical and spherical) for production of electrons

Axial dose (parallel to the main axis of the X-ray tube) and transverse dose (perpendicular to the main axis of the X-ray tube) were 0.5–10 cm from the active center of the electron generating source. These dosages were calculated and compared for miniature tubes (**Figure 2**) using the F6 tally for the energies of 40, 45, and 50 Kev. During the calculation of the axial and transverse doses using a cold carbon source (cylindrical and spherical) in all the three tubes, the average run times of programs were equal to

DOI: http://dx.doi.org/10.5772/intechopen.1001904

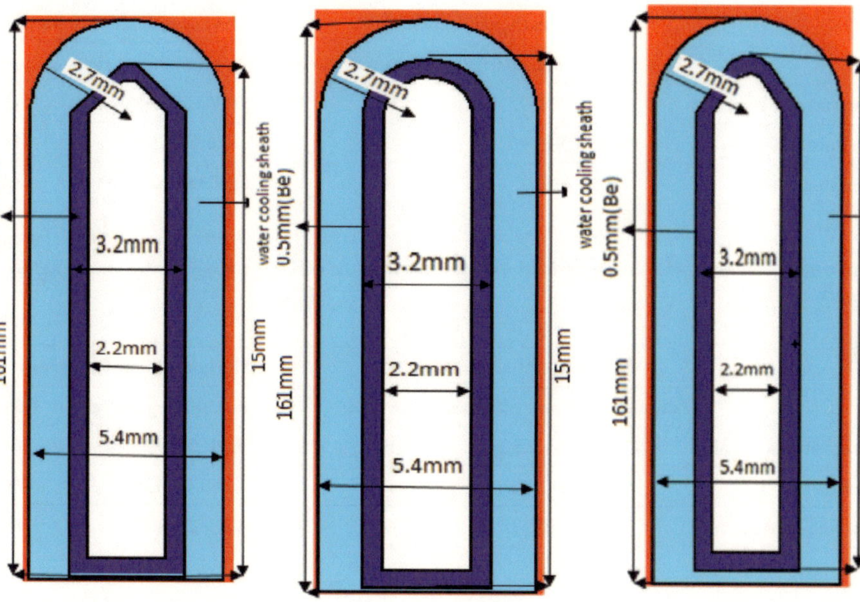

Figure 1.
A view of miniature X-ray tubes simulated for conducting the study.

Figure 2.
Simulated miniature tubes with electronic source.

	The cold cathode (carbon)		The hot cathode (tungsten)	
	Number of running particles	Run time	Number of running particles	Run time
Hemispherical-conical	2,144,007,572	3782	1,507,221,585	6000
Hemispherical	2,139,594,043	4007	1,368,439,172	6000
Truncated-conical	2,139,299,717	4022	1,359,805,034	6000

Table 1.
Comparison of the number of running particles with the tungsten and carbon cathodes during the calculation of the axial dose.

	The cold cathode (carbon)		The hot cathode (tungsten)	
	Number of running particles	Run time	Number of running particles	Run time
Hemispherical-conical	2,138,885,625	3950	1,362,548,727	6000
Hemispherical	2,142,968,016	4022	1,448,481,725	6000
Truncated-conical	2,141,088,748	4037	1,352,237,802	6000

Table 2.
Comparison of the number of running particles with the tungsten and carbon cathodes during the calculation of the transverse dose.

2670 min and 3660 min for cylindrical and spherical sources, respectively. Moreover, the average statistical error for the first 10 cells was equal to 2.5 and 3% for the cylindrical and spherical sources, respectively. Finally, for the next 10 cells, in each axis, they were equal to 4.5 and 4.7% for the cylindrical and spherical sources, respectively.

2.3 Calculation of the X-rays spectra using a cylindrical cold carbon source

The X-ray spectra of the cylindrical cold carbon cathode (r = 0.8 mm and h = 0.8 mm) were obtained for miniature X-ray tubes using the F5 ring tally at a distance of 2.5 cm from the active center of the source (the origin of the coordinates) on the longitudinal and transverse axes for 50 Kev energy (**Figure 2**). The average program run times for transverse and axial dose calculations were 2600 min and 2580 min, respectively. Moreover, the mean history of the target colloidal particles for the both axes was approximately 2×10^9.

2.4 Calculation of the axial and transverse doses using a disc-shaped source

To compare the axial and transverse doses and the running particles using hot and cold cathodes, a disk with r = 0.8 mm and thickness = 0.1 mm was evaluated in two stages: once with tungsten density (the hot cathode) and once with carbon density (the cold cathode). **Tables 1** and **2** show the results.

3. Results

3.1 Doses of hemispherical and cylindrical carbon electron sources (the cold cathode)

As shown in **Figures 3** and **4**, the axial and transverse doses were obtained for hemispherical, hemispherical-conical, and truncated-conical miniature X-ray tubes

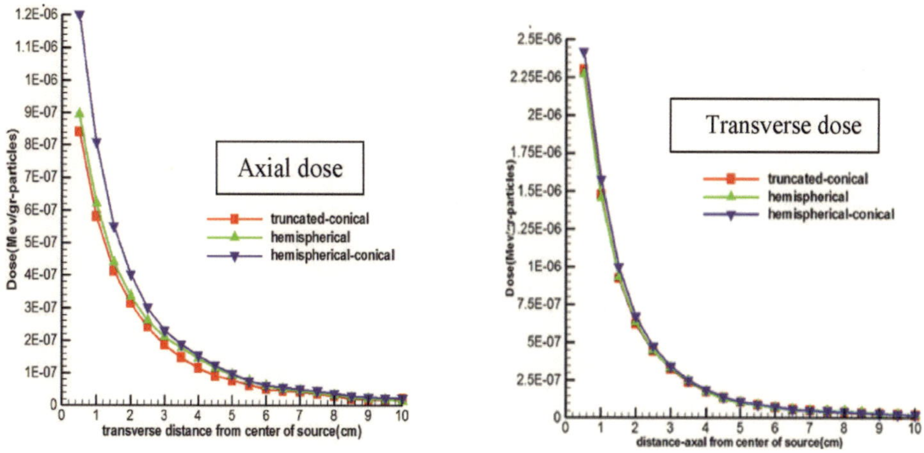

Figure 3.
The axial and transverse doses using a hemispherical cold cathode.

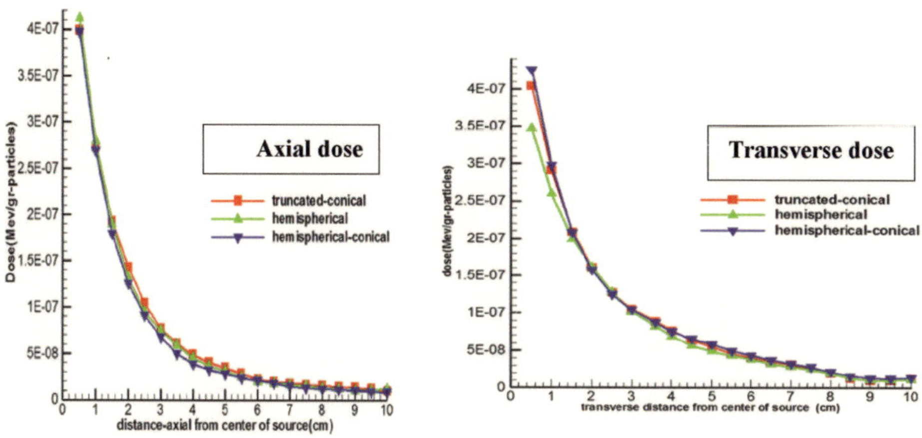

Figure 4.
The axial and transverse doses using a cylindrical cold cathode.

using 50 Kev energy in the range of 0.5–10 cm, and an electron source (the cold cathode) was selected.

3.2 Spectra of a cylindrical carbon electron source (the cold cathode)

As shown in **Figure 5**, hemispherical-conical, hemispherical, and truncated-conical X-ray spectra of miniature X-ray tubes were obtained along the longitudinal and transverse axes for 50 Kev energy at a distance of 2.5 cm from the active center of the source. Characteristic X-ray energy peaks appeared in 8.7 Kev and 10.2 Kev positions.

3.3 The axial and transverse doses of disc-shaped electron sources

The axial and transverse doses for hemispherical-conical, hemispherical, and truncated-conical miniature X-ray tubes using 50 Kev energy in the range of 0.5–10 cm by selecting two disk-shaped electron sources (the cold and hot cathodes) are shown in **Figures 6** and 7.

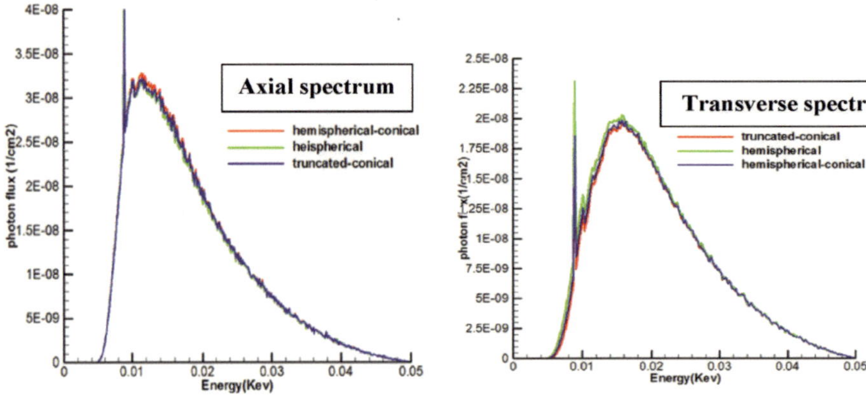

Figure 5.
Axial and transverse spectra using a cylindrical cold cathode.

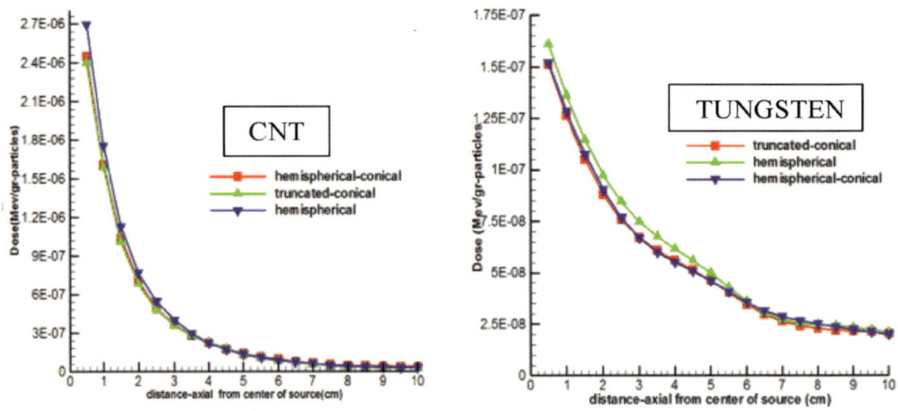

Figure 6.
The axial doses using the hot and cold cathodes.

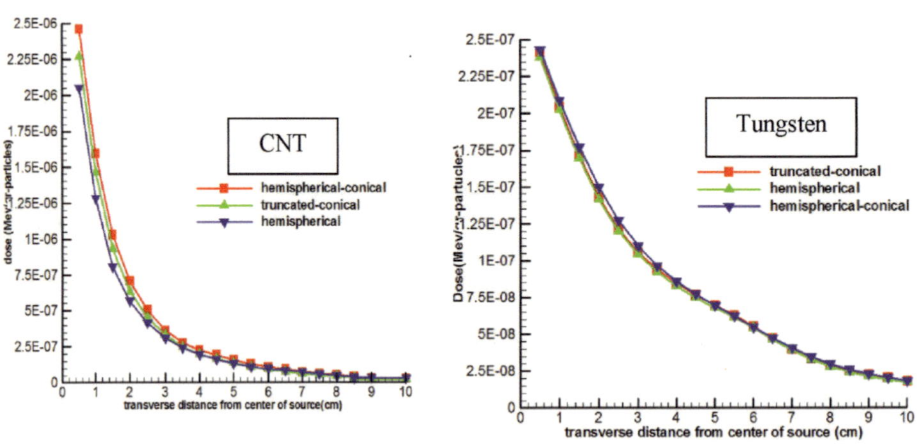

Figure 7.
The transverse doses using the hot and cold cathodes.

3.4 Running particles of a disc-shaped source

Tables 1 and **2** present the number of running particles for hemispherical-conical, hemispherical, and truncated-conical miniature X-ray tubes using disk-shaped tungsten (the hot cathode) and carbon (the cold cathode) electron sources with the same radius and thickness.

4. Discussion

4.1 Investigation of the source of carbon electron production in the form of spherical and cylindrical cold cathodes

In this section, the doses obtained from the spherical and cylindrical cold cathodes (**Figures 3** and **4**, respectively) with the same radius were compared on the longitudinal and transverse axes. As can be observed, the mean ratios of the axial and transverse doses were obtained as 5.4, 4.8, and 4.6 as well as 2.4, 2.2, and 1.8 for spherical to cylindrical cathodes in hemispherical-conical, hemispherical, and truncated-conical miniature tubes, respectively. These results showed that the spherical cold cathode was more efficient in the production of transverse and axial doses than the cylindrical cold cathode with the same radial. Further, **Figures 3** and **4** show that the hemispherical-conical X-ray tube compared to the two hemispherical and truncated-conical miniature tubes was more efficient in producing axial and transverse doses (with a spherical cold source) and transverse dose (with a cylindrical cold source). This indicated that different forms of cathode affected the output of the X-ray source in the simulation.

4.2 Analysis of the spectra of the cylindrical cold cathode

The graphs in **Figure 5** show that when using a cylindrical cold cathode, the peak of the hemispherical-conical and hemispherical graphs is higher than the rest in the transverse and axial spectra. Moreover, the larger area under the graph in the hemispherical-conical miniature tube confirmed the efficiency along the transverse axis. This means that when using a hemispherical-conical miniature x-ray tube along the transverse axis, the cylindrical cold cathode outperforms others.

4.3 Analysis of the application results of the disk-shaped carbon and tungsten electron sources

In this section, the axial and transverse doses (according to **Figure 6**) and electron production (based on **Tables 1** and **2**) were compared in terms of three hemispherical-conical, hemispherical, and truncated-conical miniature X-ray tubes. For this purpose, a disk-shaped tungsten and a carbon layer with the same thickness and radius as an electron source were used. The number of running electrons for the transverse dose of the carbon layer was 1.5, 1.47, and 1.58. Meanwhile, for the axial dose of the carbon layer, the number was 1.42, 1.56, and 1.57 times the tungsten layer, respectively, for hemispherical-conical, hemispherical, and truncated-conical X-ray tubes. In addition, the average run time of each program using carbon and tungsten layers based on **Tables 1** and **2** was 3970 and 6000 min, respectively. The mean ratio of the transverse and axial doses of the carbon source to the tungsten source was obtained to be 4.8, 4.1, and 4.4 as well as 7.4, 6.7, and 7.9 for hemispherical-conical, hemispherical, and truncated-conical tubes, respectively.

5. Conclusion

Overall, based on the results of this study, the following conclusions can be made:

The data of graphs, spectra, and tables show that the geometric structure (forms of cathode and anode and their material) affects the output of miniature X-ray tubes. In this work, this parameter had the greatest impact on the hemispherical-conical miniature X-ray tube.

The data in the tables show that the cold cathode responds in a shorter time than the hot cathode, and the number of running particles was much more with the cold cathode than with the hot cathode. In brief, the carbon cathode had a higher dose than the tungsten cathode.

According to the calculated dose, the hemispherical-conical miniature X-ray tube (similar to the tube used in the Axxent system) was in the first place, and the hemispherical miniature X-ray tube (similar to the tube used in the Intrabeam system) was in the second place.

Appendix

Appendix 1: the flow density relation

$$J_c = A_c T^2 \exp\left(\frac{-\phi}{kT}\right) \quad (1)$$

Where $J_c\left(\frac{A}{cm^2}\right)$ and $A_c = 120\frac{A}{cm^2.k}$ are constant values that are the same for all ionic heat emitters.

The temperature is in Kelvin, $K = 8.6\times10^{-5}\frac{ev}{k}$ is the Boltzmann's constant, and Φ (eV) is the cathode work function. Temperature and the work function had the greatest effect on thermionic emission, and the effect of Ac was negligible. To prevent cathode evaporation, the string temperature must have the lowest value; thus, it needs to have a small work function.

Appendix 2

The field emission of electrons from a surface to the vacuum using quantum mechanical tunneling is extremely high in response to an electric field. In general, in such a material, electrons at the Fermi surface must overcome the potential barrier applied by the work function (Φ) in the presence of an external electric field and change the potential barrier before escaping to the vacuum. In this process, the two quantities are of great importance: T (e), which is the probability of passing electrons through the barrier, and N (e), which is the source of colliding electrons. T (e) depends on the width of the barrier, which, in turn, depends on the applied field and work function of the materials. On the other hand, N (e) depends on the bond structure of the materials and the distribution of the electron at a certain temperature.

Assuming that the free-electron band structure of a metal is at 0 K temperature, the emission current follows the Fowler-Nordheim law.

$$J = \frac{A_{FN}}{\Phi t^2(y)} . E_{\text{Local}}^2 \exp\left(\frac{-B_{FN} . \Phi^{1.5}}{E_{\text{Local}}} . V(y)\right) \tag{2}$$

To extract large currents, either E_{local} must be increased or the work function (Φ) must be reduced.

Author details

Barat Barati[1*], Masumeh Kaydani[2], Fariba Farhadi Birgani[1], Mansour Zabihzadeh[3], Mohammad Javad Tahmasebi[4] and Nahid Chegeni[3]

1 Department of Radiology Technology, Shoushtar School of Medical Sciences, Shoushtar, Iran

2 Occupational Health Department, Shoushtar School of Medical Sciences, Shoushtar, Iran

3 Department of Medical Physics, School of Medicine, Ahvaz Jundishapur University of Medical Sciences, Ahvaz, Iran

4 Department of Medical Physics, School of Medicine Golestan Hospital, Ahvaz Jundishapur University of Medical Sciences, Ahvaz, Iran

*Address all correspondence to: barati943@yahoo.com

References

[1] Gazda M, Lawrence R. Principles of Radiation Therapy, Cancer Management: A Multidisciplinary Approach. 2013

[2] Sun Y. Design and Fabrication of Carbon Nanotube Array Based Field Emission Cathode for X-Ray Tube. 2013

[3] Sugie H, Tanemura M, Filip V, Iwata K, Takahashi K, Okuyama F. Carbon nanotubes as electron source in an x-ray tube. Applied Physics Letters. 2001;**78**(17):2578-2580

[4] Filip V, Filip L, Okuyama F. Miniature x-ray tubes: Current state and future prospects. Journal of Instrumentation. 2013;**8**(03):T03005

[5] Barati B, Zabihzadeh M, Birgani MJT, Chegini N, Ghahfarokhi MH, Fatahiasl J. Assessment of two hemispherical and hemispherical-conical miniature sources used in electronic brachytherapy using Monte Carlo simulation. Electronic Physician. 2017;**9**(2):3845

[6] Barati B, Zabihzadeh M, Birgani MT, Chegini N, Fatahiasl J, Mirr I. Evaluation of the effect of source geometry on the output of miniature X-ray tube for electronic brachytherapy through simulation. Journal of Biomedical Physics & Engineering. 2018;**8**(1):29

[7] McConn RJ, Gesh CJ, Pagh RT, Rucker RA, Williams III R. Compendium of Material Composition Data for Radiation Transport Modeling. Richland, WA (United States): Pacific Northwest National Lab(PNNL); 2011

[8] Collins PG, Avouris P. Nanotubes for electronics. Scientific American. 2000;**283**(6):62-69

[9] Milne W, Teo K, Mann M, Bu I, Amaratunga G, De Jonge N, et al. Carbon nanotubes as electron sources. 2006;**203**(6):1058-1063

Chapter 4

Random Sampling-and-Averaging with Variance-Reduction Techniques in Quantum Measurements

Tzu-Chien Hsueh

Abstract

This chapter reviews the technical evolution of random sampling-and-averaging (RSA) technique associated with variance reduction (VR) in the time-interval measurements of emerging microelectronic and quantum applications. First, the theoretical analysis of the RSA technique based on stochastic Monte Carlo methods is elaborated for power-efficient and highly-accurate signal and photon detections, including both synchronous and asynchronous RSA measurement techniques with superior time-domain detection resolution, scalable dynamic ranges, high linearity, high noise-immunity, and low power/area consumption. Second, to further enhance the conversion-rate of the RSA measurements, the theoretical expectations, variances, and correlation coefficients of two RSA-compatible VR techniques, self-antithetic and control-variate VRs, are comprehensively derived and expressed in the mathematical closed-forms for practical integrated-circuit (IC) hardware realization.

Keywords: antithetic variate, control variate, correlated random variable, independent and identically distributed, joint probability density function, Monte Carlo method, quantum probability amplitude, stochastic random sampling, time-correlated single-photon counting, time-domain modulo operation, time-to-digital converter, variance reduction

1. Introduction

Today, time-correlated single-photon counting (TCSPC) [1, 2] plays as the primary functionality in various emerging microelectronics and quantum technology. Different TCSPC architectures have their own pros and cons in multiple performance specifications depending on the requirements of time-to-digital conversion-rate or conversion-accuracy, which can roughly categorize TCSPC applications into two major areas as follows: (a) Quantum imaging, quantum sensing, positron emission tomography, time-resolved spectroscopy, fluorescence-lifetime imaging, light detection-and-ranging, and time-of-flight sensing, etc. employ small-area and low-power time-to-digital conversion (TDC) implementations with the disadvantage of low resolution, low accuracy, and high clock-generation power. (b) Quantum cryptography, Q-bit-state probability amplitude measurements, live-cell and tissue microscopy, and molecular imaging, etc. mainly exploit high-resolution TDC implementations with the downsides of low conversion-rates, high calibration complexity, and high-order digital filtering. In the long run of quantum-technology development, the demand for supporting both high-speed and high-resolution with low power/area consumption will be the common direction of all TCSPC realization approaches.

Therefore, this chapter reviews the theoretical analysis of random sampling-and-averaging (RSA) technique based on stochastic Monte Carlo methods used to perform high-accuracy TCSPC functionality with the downside of slow conversion rates. Then, two RSA-compatible variance-reduction (VR) techniques, self-antithetic and control-variate VRs, are elaborated in this chapter as well mainly for enhancing the conversion-rate of asynchronous RSA to realize a unified RSA-based TDC architecture for both categories of high-speed and high-resolution TCSPC applications. To verify the feasibility of the VR techniques, this chapter comprehensively summarizes the theoretical expectations, variances, and correlation coefficients in the mathematical closed-form expressions for practical integrated-circuit (IC) hardware realization.

2. Random sampling-and-averaging techniques

Monte Carlo methods have been broadly used in the fields of applied mathematics and financial engineering [3]. The main focus of this chapter, RSA measurement technique, was originated from the principle of Monte Carlo methods based on the analogy between and volume statistics and probability. In the field of stochastic processes, probability density functions (PDF) formulize the concept of probabilities to define the volumes of possible outcomes. Meanwhile, the Monte Carlo methods obtain the volumes from experiments and then interpret the volumes as probabilities. The relationship between the theoretical probability and experimental Monte Carlo method is summarized in Eqs. (1)-(3) by examining the expectation and mean values of a random variable, Y [4].

$$E[Y] = \int_{-\infty}^{\infty} y \bullet f(y) \bullet dy \tag{1}$$

$$\overline{Y} = \frac{1}{N} \bullet \sum_{n=1}^{N} Y_n \tag{2}$$

$$E[Y] = E\left[\overline{Y}\right] = \lim_{N \to \infty} \frac{1}{N} \bullet \sum_{n=1}^{N} Y_n \tag{3}$$

N is the total number of the samples; Y_n represents the n-th experimental sample of Y; E[Y] represents the expectation of the random variable Y based on its PDF, f(y); $\overline{Y}$ is equal to the mean of Y_n and the Monte Carlo estimate of the expectation E[Y]. If N increases toward infinity, $\overline{Y}$ will converge to E[Y] as well as the expectation of $\overline{Y}$, E [$\overline{Y}$]. On the condition that Y is an independent and identically distributed (I.I.D.) random variable, the variances Var[Y] and Var[$\overline{Y}$] and standard deviations σ_Y and $\sigma_{\overline{Y}}$ of Y and $\overline{Y}$ can be respectively expressed by Eqs. (4) and (5) [4].

$$Var[Y] = \sigma_Y^2 = E\left[(Y - E[Y])^2\right] = \int_{-\infty}^{\infty} (y - E[Y])^2 \bullet f(y) \bullet dy \tag{4}$$

$$\begin{aligned} Var\left[\overline{Y}\right] &= \sigma_{\overline{Y}}^2 = E\left[\left(\overline{Y} - E\left[\overline{Y}\right]\right)^2\right] = E\left[\left(\overline{Y} - E[Y]\right)^2\right] \\ &= Var\left[\frac{1}{N} \bullet \sum_{n=1}^{N} Y_n\right] = \frac{1}{N^2} \bullet \sum_{n=1}^{N} Var[Y_n] = \frac{1}{N^2} \bullet N \bullet Var[Y] = \frac{1}{N} \bullet \sigma_Y^2 \end{aligned} \tag{5}$$

Eqs. (4) and (5) provide two pieces of important information as follows: (a) Since the delta between the Monte Carlo estimate and ideal expectation, $(\overline{Y} - E[Y])$, represents the "error" of the experimental approximation, the variance and standard deviation of $\overline{Y}$ shown in Eq. (5) represent the degree of accuracy of the Monte Carlo method in estimating E[Y] with a finite sampling number N based on the Central Limit Theorem [4]. (b) The variance $Var[\overline{Y}]$, which is the power of the estimation error, decreases with the sampling number N. This is basically the Weak Law of Large Numbers (WLLN) [4], which ensures the Monte Carlo estimate $\overline{Y}$ can eventually converge to the expectation E[Y] when the sampling number N reaches toward infinity. Theoretically, the WLLN is the foundation of the Monte Carlo method and/or the RSA technique to achieve high-accuracy TDCs with small amounts of hardware cost by simply increasing the sampling number N, but obviously with the penalty of time-consumption or slow conversion-rate.

To implement the RSA-based TDC in a single-photon time-interval measurement, the TCSPC system utilizes a time-to-voltage conversion (TVC) circuit, two identical voltage-controlled delay lines (VCDL), and an edge combiner to convert the one-time captured Δt information, which is the quantity under measurement, into a periodic digital clock signal, CKτ, carrying a scaled version of Δt within each clock cycle to enable a process for an unlimited number of samples. The TVC first generates a single pulse, INT, whose pulse width equals the time difference, Δt, between the rising edges of the two pulses under the detection. Then, the INT pulse width enables an analog integrator implemented by a tunable constant current source, I_I, charging the integration and parasitic capacitors, C_I and C_P, to form a DC voltage, V_{TVC}. The time-interval, Δt, is converted and retained in the voltage domain as a differential DC voltage, $\Delta V = V_{DD} - V_{TVC} = K_{TVC} \cdot \Delta t$, where K_{TVC} is the conversion-gain of the TVC set by the magnitudes of I_I and C_I. A variable-gain amplifier (VGA) buffers the constant voltage information with its gain, K_{VGA}, to one of the following VCDLs. Because of the control voltage difference between V_{DD} and V_{VGA}, these two identical VCDLs generate two clock signals, CK_1 and CK_2, with a common frequency, 1/T, and a constant delay, $\tau = (K_{TVC} \cdot K_{VGA} \cdot K_{DL}) \cdot \Delta t$, where K_{DL} is the conversion-gain of the VCDLs. After a rising-edge combiner, the CKτ signal merged from CK_1 and CK_2 is a periodic pulse carrying the scaled time-interval information, τ, as its duty-cycle in every T. During this Δt-to-τ conversion-process, the maximum range of the time-interval measurement Δt_{MAX} is converted to the period of CKτ (T) so that the dynamic ranges for different time-interval (Δt) measurements can be set by the overall circuit conversion-factor, $(K_{TVC} \cdot K_{VGA} \cdot K_{DL})$, on the condition of a certain T.

At this point, CKτ is sampled by a single digital Data Flip-Flop (DFF), which is basically a 1-bit TDC process per sample. The sampling clock of the DFF, CK_{DCO}, is independently generated by a free-running ring oscillator digitally controlled by a pseudo-random-binary-sequence generator (PRBS) for random frequency modulation; i.e., a digital-controlled oscillator (DCO). With this random sampling frequency capability, the assumption of the uncorrelation [4, 5] between CKτ and CK_{DCO} can be guaranteed, and the waveform or voltage of CKτ at any time instant within T can have an equal chance to be sampled by CK_{DCO} to form a one-dimensional geometric probability density function [4]. Any sampled voltage from CKτ will be converted to either a Logic-1 or Logic-0 at the DFF output Y_n based on the sampled voltage larger or less than the intrinsic threshold of the DFF. After N_{DCO} sampling cycles, the accumulation of these N_{DCO} outcomes at the DFF output (Y_1, Y_2, ..., $Y_{N_{DCO}}$) can be equivalently treated as N_{DCO} times of the 1-bit TDC process all sampled within a single T due to the repeatability of CKτ; i.e., a modulo-T operation.

The RSA process can be theoretically described in different points of view as follows: (a) When the sampling number N_{DCO} is sufficiently large, the PDF of the DCO sampling instants should be uniformly distributed across the period of the CKτ, i.e., $f_{DCO}(t) = 1/T$, $t \in [0, T)$, so in the time-domain the probability of obtaining a Y_n as Logic-1, P_1, is therefore the ratio of τ to T (= τ/T). (b) In the voltage domain, the probability function of Y, f(y), is a Bernoulli distribution [4] because Y_n has only two possible outcomes, Logic-1 or Logic-0, and their corresponding probability values are P_1 (= τ/T) and P_0 (= 1 − P_1), respectively. (c) The Monte Carlo estimate, $\overline{Y}$, can be obtained by finding the ratio between the accumulated number of Logic-1s, N_Y, and the accumulated number of samples, N_{DCO}. The mathematical expressions of these three points of view are respectively summarized in the 1st, 2nd, and 3rd lines of Eq. (6) [6]:

$$
\begin{aligned}
E[Y] &= \int_0^T y(t) \bullet f_{DCO}(t) \bullet dt = \int_0^\tau 1 \bullet \frac{1}{T} \bullet dt = \frac{\tau}{T} \\
&= \int_{-\infty}^{\infty} y \bullet f(y) \bullet dy = 1 \bullet P_1 + 0 \bullet P_0 = P_1 \\
&= E\left[\overline{Y}\right] = \lim_{N_{DCO}\to\infty} \frac{\sum_{n=1}^{N_{DCO}} Y_n}{N_{DCO}} = \lim_{N_{DCO}\to\infty} \frac{N_Y}{N_{DCO}}
\end{aligned}
\tag{6}
$$

$$
\begin{aligned}
Var\left[\overline{Y}\right] &= \sigma_{\overline{Y}}^2 = \frac{1}{N_{DCO}} \bullet \sigma_Y^2 \\
&= \frac{1}{N_{DCO}} \bullet \left(E\left[Y^2\right] - E^2[Y]\right) = \frac{1}{N_{DCO}} \bullet P_1 \bullet P_0
\end{aligned}
\tag{7}
$$

In Eq. (7), the variance $Var[\overline{Y}]$ (the power of the estimation error or quantization noise) degrades with the sampling number N_{DCO} and also is a function of P_1 and P_0 according to the WLLN, which is consistent with the variance of a Bernoulli random variable. Based on Eq. (6) and overall circuit conversion-factor, the time-interval Δt under each RSA measurement can be obtained by Eq. (8) as follows [6]:

$$
\Delta t = \frac{\tau}{K_{TAC} \bullet K_{VGA} \bullet K_{DL}} \approx \frac{1}{K_{TAC} \bullet K_{VGA} \bullet K_{DL}} \bullet \frac{N_Y}{N_{DCO}} \bullet T \tag{8}
$$

As mentioned, the RSA is basically the average of total N_{DCO} of 1-bit TDC process, and its achievable accuracy is a function of the sampling number N_{DCO}. Therefore, the performance metrics of an RSA-based TDC can be reflected by the accuracy and conversion-rate of each RSA measurement result $\overline{Y}$. From the standpoint of the accuracy, the quantization error Q of the RSA-based TDC process is also a random variable, and each sample of quantization error Q_n is the delta between Y_n and expectation $E[Y] = \tau/T$. Because Y_n is either Logic-1 or Logic-0, Q_n also has two possible values $(1 - \tau/T)$ and $(0 - \tau/T)$ in its Bernoulli probability function f(q). It can be easily proven that $Var[Y] = Var[Q]$, and $Var[\overline{Y}] = Var[\overline{Q}] = P_1 \cdot P_0 / N_{DCO}$ as expressed in Eq. (7). When only the quantization noise power (i.e., $Var[\overline{Q}]$) is considered, the effective number of binary bits (ENOB) of the RSA technique can be expressed as [6]:

$$ENOB = \frac{SNR}{6.02} = \begin{cases} \dfrac{log_{10}\left(\dfrac{P_0^2}{Var\left[\overline{Y}\right]}\right)}{0.602} = \dfrac{log_{10}\left(\dfrac{P_0}{P_1} \bullet N_{DCO}\right)}{0.602}, P_1 \le \dfrac{1}{2} \\ \dfrac{log_{10}\left(\dfrac{P_1^2}{Var\left[\overline{Y}\right]}\right)}{0.602} = \dfrac{log_{10}\left(\dfrac{P_1}{P_0} \bullet N_{DCO}\right)}{0.602}, P_1 > \dfrac{1}{2} \end{cases} \tag{9}$$

In Eq. (9), the signal power is either P_1^2 or P_0^2 based on the magnitude of P_1 compared to 0.5 due to the symmetric and signal-dependent variance property of the Bernoulli distribution. Conceptually, Eq. (9) can be examined by considering $N_{DCO} = 1$, then the RSA variance ($Var[\overline{Y}] = P_1 \cdot P_0$) basically matches with the variance of a single sample experiment of flipping a coin. If the coin is fair, the probability of obtaining a single "head" or "tail" is $0.5 = P_1 = P_0$ so the SNR = 0 dB and ENOB = 0 bit based on Eq. (9). By increasing the number of the experiment N_{DCO}, the quantization error power $Var[\overline{Y}]$ exhibits a slope of −3 dBW per octave of N_{DCO}. For example, to perform an RSA-based TDC with 12-ENOB accuracy, N_{DCO} must be as large as 2^{24} (16.78 million) samples. An IC realization for this number of samples is actually not expensive at all because the RSA-based TDC only needs to accumulate 2^{24} of the 1-bit data for averaging. From the standpoint of conversion-rate, it is determined by the DCO frequency and the sampling number N_{DCO}. In the same example mentioned above, an average 4-GHz DCO sampling frequency with $N_{DCO} = 2^{24}$ requires up to 4.19 milliseconds (ms) to complete a single RSA-based TDC. In high-speed quantum applications, 4.19 ms per pixel of single-photon detection could be too slow in terms of the deadtime and frame-rate requirements. One approach of boosting the conversion-rate is to exploit the multiple inherent phases of the DCO to simultaneously sample $CK\tau$ through multiple DFFs in parallel per DCO cycle. For example, a four-stage differential DCO can linearly improve the number of sampling phases by $(4 \cdot 2) \times$ per cycle. This way can moderately increase the conversion-rate up to 1.91 kHz (8/4.19-ms) for a 12-ENOB RSA-based TDC. It is important to note that ENOB here only considers quantization error since the quantity under the RSA measurement is essentially a constant signal (DC). Any non-zero frequency (AC) noise induced jitter can be filtered out by the averaging process (i.e., integrator-based filtering); any phase-noise on sampling clock CK_{DCO} can actually improve the continuity (i.e., zero step size) of the random sampling PDF. Therefore, the IC implementation of the RSA-based TDC can possess both high noise-immunity and measurement-accuracy.

3. Synchronous random sampling-and-averaging

Synchronous RSA means the number of samples per T during each RSA sampling process is constantly set by a certain value of oversampling ratio (OSR) though the uniformly distributed PDFs of all CK_{DCO} sampling instants (or edges) are still I.I.D. Note that the definition of OSR in this chapter is defined by the number of samples per $CK\tau$ period (T), which is not the same as the OSR usually defined in the sampling theorem for anti-aliasing. Multiple examples of synchronous RSA are listed as follows: (a) OSR = 1/2, one CK_{DCO} edge uniformly samples two periods of $CK\tau$ ($2{\cdot}T$). (b) OSR = 1, one CK_{DCO} edge uniformly samples one period of $CK\tau$ (T). (c) OSR = 2, one CK_{DCO} edge uniformly samples a half period of $CK\tau$ ($0.5{\cdot}T$). (d) OSR = 4, each $CK\tau$ period is sampled by four independent CK_{DCO} edges, and each sampling edge possesses a uniform PDF bounded within one of the T/4 regions. In sum, the probability of each CK_{DCO} sampling edge occurs uniformly within the region of T/OSR. Under this assumption for synchronous RSA, though $CK\tau$ can be seamlessly and uniformly sampled by CK_{DCO} regardless of the values of OSR, however, OSR plays the key role of determining the accuracy of synchronous RSA. In reality, generating these well-bounded random sampling edges in IC implementation is challenging while the step size or resolution of the random sampling edges must be higher than the target ENOB of each synchronous RSA measurement. These realization issues can be resolved by asynchronous RSA elaborated in the next section. The assumption here is that the resolution of each CK_{DCO} sampling PDF for synchronous RSA is sufficient to maintain an almost "continuous" probability density function [4] within its own distribution region set by T/OSR. Overall, though the realization of synchronous RSA is impractical, its concept can help readers to understand the theories behind the asynchronous RSA and VR techniques described in the second half of this chapter.

Based on the definition of the synchronous RSA described above, the theoretical variances with respect to different OSRs can be derived. In the case of OSR = 1, the probability of obtaining a Logic-1 per sample is exactly $\tau/T = P_1$, so the expectation and theoretical variance of each RSA measurement are equal to the results shown in Eqs. (6) and (7), respectively. In the cases of OSR < 1 (i.e., subsampling), for example, when OSR = 1/4, the sampling region of each CK_{DCO} edge is extended to $4{\cdot}T$, the probability of obtaining a Logic-1 stays the same as $\tau/T = (4{\cdot}\tau)/(4{\cdot}T) = P_1$ since the positive duty-cycle of $CK\tau$ is also extended by 4 times. Overall, whenever OSR ≤ 1, the expectation and theoretical variance of synchronous RSA can be expressed by Eqs. (6) and (7), respectively, on the condition that OSR must be always the reciprocal of a positive integer number to satisfy a constant $CK\tau$ duty-cycle within each uniformly distributed sampling region set by T/OSR.

When OSR > 1, each uniformly distributed sampling region is labeled by an integer index k, $k \in [1, OSR]$; there are total OSR sampling outcomes per T, and only one of these sampling outcomes can possibly be Logic-1 or 0, which is the region containing the $CK\tau$ transition from high to low. For the example of OSR = 2 and $\tau/T < 0.5$, the $CK\tau$ period is equally split into two sampling regions (k = 1 to 2). Only the first region (k = 1) can possibly obtain a Logic-1 or Logic-0 outcome because $\tau/T < 0.5$, and the Logic-1 probability is $\tau/(T/2) = 2{\cdot}P_1$; note that $P_1 = \tau/T$ is the probability of obtaining Logic-1 in the cases of OSR ≤ 1. On the other hand, the outcome of the second region (k = 2) is always Logic-0 since $CK\tau$ does not transit in the region of k = 2. The equivalent outcome of each synchronous RSA measurement is the average of k = 1 and k = 2 sampling outcomes. This example shows that the variance of each synchronous RSA, $Var[\overline{Y}]$, is a function of OSR and the index k indicating the region of containing the CKτ transition from high to low. In general, the oversampling process reduces the power of quantization error of the RSA-based TDC because the regions without any CKτ transition designate the coarse information (MSBs) of the CKτ duty-cycle while the region with the CKτ transition provides the fine information (LSB). Thus, a larger value of OSRs confines the uncertain outcome due to the CKτ transition within a finer region and equivalently enhances the resolution of the synchronous RSA measurement similar to the quantization process of an analog-to-digital conversion.

In the cases of OSR ≥ 1, the theoretical variance of the synchronous RSA can be acquired by finding the total covariance sum with the joint PDFs of all samples within each T, so the expectation and theoretical variance of the synchronous RSA measurement can be generalized as follows [6]:

$$E[\overline{Y}] = \lim_{N_{DCO}\to\infty} \frac{1}{N_{DCO}} \bullet \sum_{n=1}^{N_{DCO}} Y_n = \lim_{N_{DCO}\to\infty} \frac{1}{\frac{N_{DCO}}{OSR}} \bullet \sum_{n=1}^{\frac{N_{DCO}}{OSR}} \frac{\sum_{k=1}^{OSR} Y_k}{OSR}$$

$$= \lim_{N_{DCO}\to\infty} \frac{1}{\frac{N_{DCO}}{OSR}} \bullet \sum_{n=1}^{\frac{N_{DCO}}{OSR}} Y_{OSR,n} \tag{10}$$

$$\frac{k-1}{OSR} < \left(\frac{\tau}{T} = P_1 = 1 - P_0\right) \leq \frac{k}{OSR}, k \in 1, 2, \ldots, OSR$$

$$Var[\overline{Y}] = Var[\overline{Q}] = \sigma_{\overline{Y}}^2 = \begin{cases} \dfrac{(OSR \bullet P_1 - k + 1)}{N_{DCO}} \bullet \left(\dfrac{k}{OSR} - P_1\right), & \dfrac{k}{OSR} \leq \dfrac{1}{2} \\ \dfrac{(OSR \bullet P_0 + k - OSR)}{N_{DCO}} \bullet \left(\dfrac{OSR - k + 1}{OSR} - P_0\right), & \dfrac{k}{OSR} > \dfrac{1}{2} \end{cases} \tag{11}$$

Based on Eq. (11), the correlations induced by the oversampling process reduce the overall variances. That is, evenly splitting each T by OSR and taking the average of all sub-region outcomes to reach the overall result of each synchronous RSA measurement are actually producing negative correlations among all the samples. Note that "k" in Eq. (10) simply represents the index of the summation operator to find $Y_{OSR,n}$ per T, but "k" in Eq. (11) is a specific integer number within 1 to OSR based upon the transition of CKτ; i.e., for a certain τ/T under the RSA measurement, only a certain k can meet the 1st line of Eq. (11). Also, note that, regardless of the value of OSR, the WLLN is always valid.

4. Asynchronous random sampling-and-averaging

The asynchronous RSA technique can be practically realized by power and area efficient ICs and perform I.I.D. random sampling within the period of CKτ without strict sampling PDF boundaries and frequency relationships between CKτ and CK_{DCO}. Before explaining how I.I.D. random sampling can be in an asynchronous manner for practical RSA realization, some parameters are defined as follows: (a) $T_{DCO,n}$ is the n-th period of the DCO. (b) $t_{SAMP,n}$ is the n-th absolute time at the n-th sampling edge of the DCO. (c) $\Delta T_{PRBS,n}$ is the n-th DCO period dynamically extended by the PRBS generator ($T_{DCO,n} = T_{DCO,MIN} + \Delta T_{PRBS,n}$) between ($T_{DCO,MIN}$ + 0) and ($T_{DCO,MIN} + \Delta T_{PRBS,MAX}$) [7]. Both $T_{DCO,MIN}$ and $\Delta T_{PRBS,MAX}$ can be coarsely adjusted by their own static controls in the DCO circuit. (d) For the sake of simplicity, $CK\tau$ and CK_{DCO} are assumed to exhibit coincident rising edges at t = 0.

Though naturally the DCO can perform a phase-noise accumulation under the presence of device Johnson (thermal) noise, flicker noise, and power-supply noise, a strong phase-noise accumulation is required to form a widely distributed random sampling PDF. Therefore, a noise-energy dominated PRBS noise source is intentionally added in the DCO for the asynchronous RSA technique. Theoretically, the combinational effect of multiple I.I.D. noise sources are all applicable to the analysis result in this section, but the asynchronous RSA technique relies on artificial PRBS noise to effectively scramble and avoid the synchronicity between $CK\tau$ and CK_{DCO}. Meanwhile, although the circuit/device noise sources exhibit much lower energy than the PRBS can, they offer arbitrarily small phase-noise accumulations to fill the gaps among discrete PRBS noise step sizes. Therefore, the combinational phase-noise accumulations from the circuits/devices and PRBS generator can help CK_{DCO} to possess an almost continuous sampling PDF.

Under the parameters and noise definitions described above, the n-th absolute sampling time can be represented by Eq. (12) as follows [6]:

$$\begin{aligned} t_{SAMP,n} &= \sum_{k=1}^{n} T_{DCO,k} = \sum_{k=1}^{n} (T_{DCO,MIN} + \Delta T_{PRBS,k}) \\ &= n \bullet T_{DCO,MIN} + \sum_{k=1}^{n} \Delta T_{PRBS,k} \end{aligned} \tag{12}$$

Each $t_{SAMP,n}$ consists of two components: the deterministic term ($n \cdot T_{DCO,MIN}$) and stochastic term created by the phase-noise accumulation, which represents the randomness of each CK_{DCO} sampling instant with a PDF, $f_{DCO,n}(t)$, so the occurrence of the n-th CK_{DCO} sampling edge is a random variable having its own PDF, $f_{DCO,n}(t)$, with the density magnitude across the distribution span. Also, as shown in Eq. (12), the stochastic term is the accumulation of "n" I.I.D. random variables ($\Delta T_{PRBS,k}$, k = 1 to n) produced by the PRBS generator over "n" times, so $f_{DCO,n}(t)$ is the convolution result of "n" uniformly distributed PDFs from the PRBS generator. For example, if n = 1, the stochastic term has one random variable from the PRBS generator with the PDF, $f_{DCO,1}(t)$, which is the fundamental uniform distribution with a distribution span over $\Delta T_{PRBS,MAX}$ and constant density magnitude of $1/\Delta T_{PRBS,MAX}$. If n = 2, the stochastic term is the accumulation of two I.I.D. random variables, which both are I.I.D. but sequentially produced by the PRBS generator. Therefore, $f_{DCO,2}(t)$ is the convolution of the two individual uniformly distributed PDFs because of the Convolution Theorem [4], and it exhibits an isosceles triangular distribution with a distribution span across $2 \cdot \Delta T_{PRBS,MAX}$ and $1/\Delta T_{PRBS,MAX}$ peak density magnitude. By increasing the of number of samples, the Central Limit Theorem guarantees that the PDF of the stochastic term converges to a Gaussian distribution independent from the PDF of each random variable produced by the PRBS generator. That is, when $n \gg 1$, the mean, standard deviation, span, and peak of $f_{DCO,n}(t)$ are converged to the expressions as follows [6]:

$$Mean_n = n \bullet T_{DCO,MIN} + \frac{n}{2} \bullet \Delta T_{PRBS,MAX},\ Span_n = n \bullet \Delta T_{PRBS,MAX}$$

$$STD_n = \sqrt{\frac{n}{12}} \bullet \Delta T_{PRBS,MAX},\ Peak_n = \sqrt{\frac{6}{\pi \bullet n}} \bullet \frac{1}{\Delta T_{PRBS,MAX}}$$

$$
\begin{aligned}
f_{DCO,n}(t) &= Conv\left[f_{DCO,n-1}(t), f_{DCO,1}(t)\right], n>1 \\
&\approx \frac{1}{STD_n \bullet \sqrt{2 \bullet \pi}} \bullet exp\left[\frac{-(t-Mean_n)^2}{2 \bullet STD_n^2}\right], n \gg 1
\end{aligned}
\tag{13}
$$

Note that the time-domain variable "t" represents absolute time values referenced to t = 0. Since the deterministic term ($n \cdot T_{DCO,MIN}$) of $t_{SAMP,n}$ sets the left distribution bound of $f_{DCO,n}(t)$, the PDFs of adjacent samples quickly exhibit a large amount of distribution overlaps along with the growth of $Span_n$. Note that the overlaps among sampling PDFs in the time-domain do not change the monotonic ascending order of the $t_{SAMP,n}$ occurrences; this just indicates the correlation among the $f_{DCO,n}(t)$ due to the DCO phase-noise accumulation started from the 1st sample to the n-th sample as expressed in Eq. (12); e.g., $t_{SAMP,n-1}$ always occurs before $t_{SAMP,n}$ although the PDFs of their stochastic term, $f_{DCO,n-1}(t)$ and $f_{DCO,n}(t)$, overlap on the top of each other in the absolute time domain.

The DCO sampling PDFs generated by the approach described so far does not seem uniform and independent because: (a) when $n \gg 1$, $f_{DCO,n}(t)$ in Eq. (13) becomes a Gaussian distribution PDF; (b) all the sampling instants, $t_{SAMP,n}$, are correlated because of the DCO phase-noise accumulation. These two concerns can be actually resolved by properly setting up the noise-energy ($\Delta T_{PRBS,MAX}$) of the PRBS generator with the modulo-T operation because of the periodicity of CKτ, so the Gaussian distribution and correlated characteristics of the DCO sampling PDF, $f_{DCO,n}(t)$, can be turned into the uniform distribution and independent characteristic of modulo-T sampling PDFs, $f_n(t)$. The modulo-T operation is automatically achieved when converting Δt into the duty-cycle τ/T of the periodic signal CKτ. The detail of the modulo-T operation is described below.

If $n \gg 1$, the Gaussian distribution span of $f_{DCO,n}(t)$ covers several CKτ periods, and the CKτ period T equivalently slices the entire Gaussian PDF into several pieces in the absolute time-domain. That is, these sliced PDFs maintain their own magnitude distributions but are all bounded within the same distribution span T, and all sample the identical CKτ waveform $y_n(t)$, $t \in [0, T)$. In other words, the distributions of all sliced PDFs are strictly confined within a modulo-T time-interval between 0 and T. Thus, the net density-magnitude at any time instant within [0, T) is the linear summation of all sliced PDFs; i.e., the equivalent PDF of the n-th DCO sampling instant $f_n(t)$ is the summation of all sliced PDFs from $f_{DCO,n}(t)$ [6]:

$$
\begin{aligned}
S &= Ceil\left[\frac{1}{2} \bullet \frac{Span_n}{T}\right], Mean_n' = Mod[Mean_n, T] \\
f_n(t) &\approx \frac{1}{STD_n \bullet \sqrt{2 \bullet \pi}} \bullet \sum_{k=-S}^{S} exp\left[\frac{-(t-Mean_n'-k \bullet T)^2}{2 \bullet STD_n^2}\right], \quad n \gg 1
\end{aligned}
\tag{14}
$$

where the time-domain variable "t" is confined within [0, T); (2·S + 1) is the number of segments set by the $Span_n$; $Mean_n$' and k·T are used to shift these (2·S + 1) segments to the modulo-T time-interval, [0, T). When $n \gg 1$, $f_n(t)$ converges to a constant 1/T across the single T span; this fact can be proven by both mathematical calculation of Eq. (14) and statistical simulations. More importantly, by increasing "n", all sampling PDFs converge to a uniformly distributed PDF with a constant density magnitude 1/T across the [0, T) distribution span, independent from the parameters of $T_{DCO,MIN}$, $\Delta T_{PRBS,MAX}$, and even "n" when $n \gg 1$. In other words, for all "n" $\gg 1$, $f_n(t)$ becomes an "identically distributed" PDF, which satisfies the "second" criterion of the I.I.D. random variable and can be implemented by low-cost circuitry. This convergence of the uniform distribution also guarantees the convergence of the asynchronous RSA measurement result, which exactly matches to the expectation in Eq. (6), as follows [6]:

$$\begin{aligned} E[\overline{Y}] &= E\left[\frac{1}{N_{DCO}} \bullet \sum_{n=1}^{N_{DCO}} Y_n\right] = \frac{1}{N_{DCO}} \bullet \sum_{n=1}^{N_{DCO}} E[Y_n] = E[Y_n] \\ &= \int_0^T y_n(t) \bullet f_n(t) \bullet dt = \int_0^\tau \frac{1}{T} \bullet dt + \int_\tau^T \frac{0}{T} \bullet dt = \frac{\tau}{T} = P_1 \end{aligned} \tag{15}$$

If the value of "n" is not large enough, the modulo-T PDF may behave like a non-uniform distribution; e.g., when $\Delta T_{PRBS,MAX} = 0.25 \cdot T$, $f_n(t)$ does not perform like a uniformly distributed PDF until n = 64. However, these uniformly distributed modulo-T PDFs will not affect the overall asynchronous RSA measurement result shown in Eq. (15) since these non-uniform modulo-T PDFs can occur only when all the RSA experiments have the identical absolute time reference at t = 0 as the assumption for Eq. (12). In realistic hardware implementation, the initial conditions of $CK\tau$ and CK_{DCO} will be inevitably randomized during the sampling process so that $f_n(t)$ for all "n" can actually form uniformly distributed PDFs within [0, T) although the value of "n" is a small number. This effect can be observed by merging a naturally randomized initial condition t_{INT} into the unit impulse $\delta(t)$ expressed in Eq. (16).

Based on the Convolution Theorem [4], $f_n(t)$ can be expressed in the format similar to $f_{DCO,n}(t)$ in Eq. (13), but note that $f_n(t)$ shall follow the Circular Convolution Theorem [8] due to the combination of the modulo-T operation and linear convolution [6]:

$$\begin{aligned} f_1(t) &= CConv\left[f_{DCO,1}(t), \delta(t), T\right] \\ f_n(t) &= CConv\left[f_{n-1}(t), f_1(t), T\right], n > 1 \end{aligned} \tag{16}$$

In Eq. (16), $f_1(t)$, $f_{n-1}(t)$, and $f_n(t)$ is in the modulo-T time-domain $t \in [0, T)$; $f_{DCO,1}(t)$ and $\delta(t)$ is in the absolute time-domain $t = [0, \infty)$. It is important to emphasize that $f_1(t)$ is not only the PDF of the 1st sampling instant but also the elementary PDF to derive $f_n(t)$ from $f_{n-1}(t)$ based on the Circular Convolution Theorem.

Several important attributes of $f_1(t)$ are summarized as follows: The distribution of $f_1(t)$ constantly starts at the remainder of $T_{DCO,min}/T$. Whenever the PDF reaches t = T, it circulates back to t = 0 and then continues its distribution toward t = T. In the first case of $\Delta T_{PRBS,MAX} < T$, for example, $\Delta T_{PRBS,MAX} = 0.25 \cdot T$, intuitively $f_1(t)$ has non-zero values from a certain "t" to "$t + 0.25 \cdot T$" within [0, T). In the second case of $\Delta T_{PRBS,MAX} = T$ or $Mod[\Delta T_{PRBS,MAX}, T] = 0$, $f_1(t)$ circulates multiple integer cycles uniformly from a certain "t" within [0, T) and then back to the same "t". In the third case of $\Delta T_{PRBS,MAX} > T$ and $Mod[\Delta T_{PRBS,MAX}, T] \neq 0$, $f_1(t)$ exhibits two non-zero density magnitudes because $f_{DCO,1}(t)$ circulates within [0, T) multiple times with a non-zero remainder, and the delta between these two density magnitudes is $1/\Delta T_{PRBS,MAX}$. The third case indicates $f_1(t)$ can converge to a uniform distribution by itself when $\Delta T_{PRBS,MAX} \gg T$. The attributes of $f_1(t)$ seem insignificant since $f_n(t)$ will converge to a uniform distribution anyway; however, although all $f_n(t)$ are identical when $n \gg 1$, this elementary PDF, $f_1(t)$, plays an important role from the standpoint of the correlations among all sampling PDFs, $f_n(t)$, n = 1 to N_{DCO}.

To elaborate the correlations among all sampling PDFs, first of all, the covariance between adjacent samples, Y_n and Y_{n+1}, can be examined as follows convolution [6]:

$$\begin{aligned}
Cov[Y_n, Y_{n+1}] &= \iint_R (y_n - E[Y]) \bullet (y_{n+1} - E[Y]) \bullet f(y_n, y_{n+1}) \bullet dy_n \bullet dy_{n+1} \\
&= \int_0^T (y_n(t) - P_1) \bullet (y_{n+1}(t) - P_1) \bullet f_{n,n+1}(t) \bullet dt \\
&= (1-P_1) \bullet (1-P_1) \bullet \int_0^\tau \left(f_{n,n+1}(t) \middle| \begin{matrix} Y_n = 1 \\ Y_{n+1} = 1 \end{matrix} \right) \bullet dt \\
&\quad + (1-P_1) \bullet (0-P_1) \bullet \int_\tau^T \left(f_{n,n+1}(t) \middle| \begin{matrix} Y_n = 1 \\ Y_{n+1} = 0 \end{matrix} \right) \bullet dt \\
&\quad + (0-P_1) \bullet (1-P_1) \bullet \int_0^\tau \left(f_{n,n+1}(t) \middle| \begin{matrix} Y_n = 0 \\ Y_{n+1} = 1 \end{matrix} \right) \bullet dt \\
&\quad + (0-P_1) \bullet (0-P_1) \bullet \int_\tau^T \left(f_{n,n+1}(t) \middle| \begin{matrix} Y_n = 0 \\ Y_{n+1} = 0 \end{matrix} \right) \bullet dt
\end{aligned} \tag{17}$$

where the 1st line of Eq. (17) is based on the fundamental covariance definition of two random variables, Y_n and Y_{n+1}, on the same sample space, R, with their joint PDF, $f(y_n, y_{n+1})$, and PDF variables, y_n and y_{n+1}. The 2nd line of Eq. (17) represents the covariance in the format of one-dimensional geometric probability with the modulo-T time-domain PDF variables, $y_n(t)$ and $y_{n+1}(t)$, and joint PDF, $f_{n,n+1}(t)$. $y_n(t)$ and $y_{n+1}(t)$ are the same because of the CKτ periodicity, and the possible outcomes of Y_n and Y_{n+1} are either Logic-1 or Logic-0 both with the expectation $E[Y] = \tau/T = P_1$ verified in Eq. (15). By taking advantage of simple binary values, $Cov[Y_n, Y_{n+1}]$ can be expanded into the summation of four conditional covariances based on the total four possible combinations of Y_n and Y_{n+1} with their corresponding conditional joint PDFs as shown in the 3rd to 6th lines of Eq. (17). In general, $f_n(t)$ is a uniform distribution for all $\Delta T_{PRBS,MAX}$ scenarios when $n \gg 1$, so $f_n(t)|Y_n = 1$ and $f_n(t)|Y_n = 0$ (the conditional PDFs) are set by $y_n(t)$ across the modulo-T time interval. To further reach the conditional joint PDFs, $f_{n,n+1}(t)|Y_n$, each $f_n(t)|Y_n$ has to circularly convolute with the fundamental PDF element, $f_1(t)$, which is a function of $\Delta T_{PRBS,MAX}$, so different $f_1(t)$ generate their corresponding $f_{n,n+1}(t)|Y_n = 1$ and $f_{n,n+1}(t)|Y_n = 0$ as follows [6]:

$$\begin{aligned}
\left(f_{n,n+1}(t)|Y_n = 1\right) &= CConv\left[(f_n(t)|Y_n = 1), f_1(t), T\right] \\
\left(f_{n,n+1}(t)|Y_n = 0\right) &= CConv\left[(f_n(t)|Y_n = 0), f_1(t), T\right]
\end{aligned} \tag{18}$$

Note that $f_n(t)$ in Eq. (16) and $f_{n,n+1}(t)|Y_n$ in Eq. (18) are different since $f_n(t)$ is calculated from $f_{n-1}(t)$ and becomes independent from $f_1(t)$, but $f_{n,n+1}(t)|Y_n$ is obtained from $f_n(t)|Y_n$, and $f_1(t)$ determines their correlation. Eventually, the conditional joint PDFs of each scenario are derived from $y_{n+1}(t)$ across $t \in [0, T)$ to thoroughly include all possible conditions of Y_n and Y_{n+1}.

When $Mod[\Delta T_{PRBS,MAX}, T] = 0$, the four conditional joint PDFs all maintain constant density magnitudes within their own integral time-intervals, $[0, \tau)$ and $[\tau, T)$, so the covariance of the adjacent samples can be further derived for this scenario easily [6]:

$$If\ Mod[\Delta T_{PRBS,MAX}, T] = 0,$$

$$Cov[Y_n, Y_{n+1}] = P_0 \bullet P_0 \bullet \frac{P_1}{T} \bullet \tau - P_0 \bullet P_1 \bullet \frac{P_1}{T} \bullet (T - \tau) \tag{19}$$

$$-P_1 \bullet P_0 \bullet \frac{P_0}{T} \bullet \tau + P_1 \bullet P_1 \bullet \frac{P_0}{T} \bullet (T - \tau) = 0$$

Based on the result of Eq. (19), i.e., a zero covariance between any adjacent samples Y_n and Y_{n+1}, and the accumulated relation from $f_1(t)$ to $f_n(t)$ shown in Eq. (16), $Mod[\Delta T_{PRBS,MAX}, T] = 0$ is the necessary condition for all, not just adjacent, $f_n(t)$ to be "independent." By consolidating the identicality and independency of $f_n(t)$ for all $\Delta T_{PRBS,MAX}$ scenarios, $Mod[\Delta T_{PRBS,MAX}, T] = 0$ is the fundamental criterion to perform an asynchronous RSA-based TDC with I.I.D. random sampling PDFs. This result matches the requirement of a synchronous RSA-based TDC when OSR = 1 and the CK_{DCO} sampling PDF is uniformly distributed across one $CK\tau$ cycle, T. The main difference between the two is that the asynchronous RSA-based TDC requires more time per sample due to the deterministic offset $T_{DCO,MIN}$. However, this deterministic offset per sample offers a margin for practical circuit implementation in the asynchronous RSA-based TDC.

Accurately performing "$Mod[\Delta T_{PRBS,MAX}, T] = 0$" may increase the cost of asynchronous RSA-based TDC implementation. To resolve this issue, the case of $\Delta T_{PRBS,MAX} > T$ could be considered though the result of Eq. (19) indicates that the non-uniform conditional joint PDFs induce non-zero covariances. These non-uniform conditional joint PDFs are mainly caused by the non-uniform $f_1(t)$; however, the distribution of $f_1(t)$ can be flattened by setting $\Delta T_{PRBS,MAX} \gg T$ so that all the conditional joint PDFs can be roughly uniform. In other words, satisfying $\Delta T_{PRBS,MAX} \gg T$ can be easily implemented without the exact relationship between $\Delta T_{PRBS,MAX}$ and T [7, 9], but the downside is to make conversion-rate even slower since $T_{DCO,MIN}$ is inevitably enlarged with $\Delta T_{PRBS,MAX}$ when $\Delta T_{PRBS,MAX} \gg T$. The compromise between the hardware cost and conversion-rate further confirms the necessity of variance-reduction (VR) techniques discussed in the following sections.

5. Self-antithetic variance reduction

The analysis of self-antithetic variance reduction (SAVR) technique in an asynchronous RSA system can be started from formulating the variance of a Monte Carlo estimate in general [10, 11]:

$$
\begin{aligned}
Var\left[\overline{Y}\right] &= Var\left[\frac{\sum_{n=1}^{N_{DCO}} Y_n}{N_{DCO}}\right] = \frac{Var\left[\sum_{n=1}^{N_{DCO}} Y_n\right]}{N_{DCO}^2} = \frac{\sum_{n=1}^{N_{DCO}}\sum_{k=1}^{N_{DCO}} Cov[Y_n, Y_k]}{N_{DCO}^2} = \frac{E\left[\sum_{k=1}^{N_{DCO}} Cov[Y_n, Y_k]\right]}{N_{DCO}} \\
&= \frac{\sum_{n=1}^{N_{DCO}} Var[Y_n]}{N_{DCO}^2} + \frac{2\bullet\sum_{n=2}^{N_{DCO}}\sum_{k=1}^{n-1} Cov[Y_n, Y_k]}{N_{DCO}^2} \\
&= \frac{P_1\bullet P_0}{N_{DCO}} + \frac{2\bullet\sum_{n=2}^{N_{DCO}}\sum_{k=1}^{n-1} Cov[Y_n, Y_k]}{N_{DCO}^2} \\
&= \begin{cases} \dfrac{P_1\bullet P_0}{N_{DCO}} = \dfrac{Var[Y]}{N_{DCO}}, if \displaystyle\sum_{n=2}^{N_{DCO}}\sum_{k=1}^{n-1} Cov[Y_n, Y_k] = 0 \\ Var\left[\overline{Y}_{SA}\right] < \dfrac{Var[Y]}{N_{DCO}}, if \displaystyle\sum_{n=2}^{N_{DCO}}\sum_{k=1}^{n-1} Cov[Y_n, Y_k] < 0 \end{cases}
\end{aligned}
\tag{20}
$$

In Eq. (20), "n" and "k" are the sample indexes; $Cov[Y_n, Y_k]$ represents the pairwise covariance of any two samples Y_n and Y_k. If $n = k$, then $Cov[Y_n, Y_k] = Var[Y_n]$ which is the variance of the n-th sample. If $n \neq k$, $Cov[Y_n, Y_k] = Cov[Y_k, Y_n]$ which is a symmetric covariance matrix. When Y_1, Y_2, … , and $Y_{N_{DCO}}$ are all pairwise independent, then $Cov[Y_n, Y_k] = 0$ for all $n \neq k$ in Eq. (20), and the pairwise covariance sum (the 2nd term in the 3rd line of Eq. (20)) is zero, so Eq. (20) becomes the I.I.D. situation as shown in Eq. (7). On the other hand, if there are correlations among the N_{DCO} samples of the random variable Y, and the pairwise covariance sum of all $n \neq k$ is negative (the 2nd term in the 3rd line of Eq. (20)), now the variance can be smaller than that of the I.I.D. situation, i.e., $Var[\overline{Y}_{SA}] < P_1{\cdot}P_0/N_{DCO}$, which is the main idea of SAVR. In other words, SAVR intentionally creates non-zero correlations among the samples, i.e., Y_1, Y_2, … , and $Y_{N_{DCO}}$, of the random variable Y, so that the pairwise covariance sum can be negative, and then the overall variance $Var[\overline{Y}_{SA}]$ per RSA measurement can be effectively less than the variance of the I.I.D. situation in Eq. (7).

According to the analysis about the covariances of the adjacent samples, $Cov[Y_n, Y_{n+1}]$, under three $\Delta T_{PRBS,MAX}$ scenarios (i.e., $\Delta T_{PRBS,MAX}$ <, =, or > T) in Eq. (17), the conclusion shows $Cov[Y_n, Y_{n+1}]$ can have pronounced non-zero values when Mod $[\Delta T_{PRBS,MAX}, T] < T$ due to the non-uniform conditional joint PDFs within their own integral time-intervals, $[0, \tau)$ and $[\tau, T)$. In this section, the case of non-zero covariances between adjacent samples is further extended to the deduction of any pairwise correlation between Y_n and Y_k due to the fact of the DCO phase-noise accumulation property reflected by Eqs. (12) and (16). Thus, if the technique can take advantage of these non-zero covariances, $Cov[Y_n, Y_k]$, and assure the pairwise covariance sum is negative, then VR can be successfully performed.

To find $Cov[Y_n, Y_k]$, all conditional joint PDFs of any $[Y_n, Y_k]$ pair have to be first formulated, where n = 1 to $(N_{DCO} - 1)$ and k = (n+1) to N_{DCO} are sufficient to cover all covariance elements in the symmetric covariance matrix. Though there are only four possible binary combinational outcomes of a certain $[Y_n, Y_k]$ pair, its conditional joint PDFs shall also cover all possible binary combinational outcomes from Y_{n+1} to Y_{k-1} because the accumulation property and Convolution Theorem described in Eq. (16)

form the chain of correlations from each specific Y_n through Y_{n+1}, Y_{n+2}, … , Y_{k-2}, and Y_{k-1} all the way to each specific Y_k. Thus, the conditional joint PDFs of a $[Y_n, Y_k]$ pair under all possible conditions of $[Y_n, \dots, Y_{k-1}]$ shall be generalized as follows [10]:

$$f_{n,k}(t)\Big|_{[Y_n,\dots,Y_{k-1}]=D2B[q]} = CConv\left[f_{n,k-1}(t)\Big|_{[Y_n,\dots,Y_{k-1}]=D2B[q]}, f_1(t), T\right] \tag{21}$$

where q = 0, 1, 2, … , $(2^{k-n}-1)$; all possible binary combinational conditions of $[Y_n, \dots, Y_{k-1}]$ are represented by the corresponding decimal numbers, "q", and decimal-to-binary operators, D2B[·], for the sake of simplicity. Since the location of "τ" within the modulo-T time-interval, [0, T), defines the PDF boundaries for the probability of Y_k to be Logic-1 or Logic-0, the conditional joint PDF of a $[Y_n, Y_k]$ pair under all possible conditions of $[Y_n, \dots, Y_k]$ can be obtained by forcing one part, i.e., [0, τ) or [τ, T), of the PDF in Eq. (21) to zero to satisfy one of the possible Y_k conditions [10]:

$$f_{n,k}(t)\Big|^{[Y_k]=1}_{[Y_n,\dots,Y_{k-1}]=D2B[q]} = \begin{cases} f_{n,k}(t)\Big|_{[Y_n,\dots,Y_{k-1}]=D2B[q]}, & 0 \le t < \tau \\ 0, & \tau \le t < T \end{cases} \tag{22}$$

$$f_{n,k}(t)\Big|^{[Y_k]=0}_{[Y_n,\dots,Y_{k-1}]=D2B[q]} = \begin{cases} 0, & 0 \le t < \tau \\ f_{n,k}(t)\Big|_{[Y_n,\dots,Y_{k-1}]=D2B[q]}, & \tau \le t < T \end{cases} \tag{23}$$

In other words, each conditional joint PDF under the conditions of $[Y_n, \dots, Y_{k-1}]$ in Eq. (21) can diversify into two conditional joint PDFs under the conditions of $[Y_n, \dots, Y_k]$ as shown in Eqs. (22) and (23), respectively. Note that the distribution profiles of all conditional joint PDFs are convoluted functions of τ and $f_1(t)$, and $f_1(t)$ itself is a function of $Mod[T_{DCO,MIN}, T]$ and $\Delta T_{PRBS,MAX}$. Therefore, the joint PDFs can be very different if the values of τ, $Mod[T_{DCO,MIN}, T]$, and $\Delta T_{PRBS,MAX}$ are set differently; i.e., these parameters dominantly affect the behaviors of the pairwise covariances, Cov $[Y_n, Y_k]$, which now can be generalized as follows based on the information of all conditional joint PDF of a $[Y_n, Y_k]$ pair from Eqs. (22) and (23) [10]:

$$\begin{aligned} Cov[Y_n, Y_k] &= \int_0^T (y_n(t) - P_1)\bullet(y_k(t) - P_1)\bullet f_{n,k}(t)\bullet dt \\ &= (1-P_1)^2\bullet \sum_{q=0}^{(2^{k-n-1}-1)} \int_0^{\tau} f_{n,k}(t)\Big|^{[Y_n,Y_k]=[1,1]}_{[Y_{n+1},\dots,Y_{k-1}]=D2B[q]} dt \\ &\quad -(P_1 - P_1^2)\bullet \sum_{q=0}^{(2^{k-n-1}-1)} \int_{\tau}^{T} f_{n,k}(t)\Big|^{[Y_n,Y_k]=[1,0]}_{[Y_{n+1},\dots,Y_{k-1}]=D2B[q]} dt \\ &\quad -(P_1 - P_1^2)\bullet \sum_{q=0}^{(2^{k-n-1}-1)} \int_0^{\tau} f_{n,k}(t)\Big|^{[Y_n,Y_k]=[0,1]}_{[Y_{n+1},\dots,Y_{k-1}]=D2B[q]} dt \\ &\quad +(0-P_1)^2\bullet \sum_{q=0}^{(2^{k-n-1}-1)} \int_{\tau}^{T} f_{n,k}(t)\Big|^{[Y_n,Y_k]=[0,0]}_{[Y_{n+1},\dots,Y_{k-1}]=D2B[q]} dt \end{aligned} \tag{24}$$

As shown in Eq. (24), $Cov[Y_n, Y_k]$ can be grouped into four terms based on the binary combinational outcomes of a $[Y_n, Y_k]$ pair, and each term is the summation of 2^{k-n-1} integrals of the conditional joint PDFs under all possible conditions of $[Y_{n+1}, \ldots, Y_{k-1}]$. Thus, the total number of summation terms or conditional joint PDFs in Eq. (24) is $2^2 \cdot 2^{k-n-1} = 2^{k-n+1}$ for each $Cov[Y_n, Y_k]$. In addition, each conditional joint PDF, $f_{n,k}(t)|_{[Y_n, \ldots, Y_{k-1}]}$, in Eq. (24) is obtained by $(k - n)$ times of the modulo-T circular convolutions. Overall, the computation effort of Eq. (24) is dominated by total $(k - n) \cdot 2^{k-n+1}$ modulo-T circular convolutions for each $Cov[Y_n, Y_k]$. On top of this, an overall variance, $Var[\overline{Y}]$, calculation, requires total $(N_{DCO}^2 - N_{DCO})/2$ of $Cov[Y_n, Y_k]$ elements in the whole covariance summation as shown in 3rd line of Eq. (20), and usually N_{DCO} is in the range of 2^8 to 2^{18} even with the VR techniques. In short, the computation effort of the theoretical variance with non-zero pairwise covariance formulated by Eq. (24) turns out to be impractical for the performance estimation of the RSA with SAVR technique. This issue reflects the necessity of developing a computation-efficient approach for the overall variance calculation.

To simplify the analysis process, the mechanism of SAVR is elaborated by specific examples in an asynchronous RSA system, and then general cases can be further summarized. The following two examples have common parameter setups: the time-domain quantity under the RSA measurements, $\tau/T = 0.5$ $(= P_1)$, $N_{DCO} = 256$, and $Mod[T_{DCO,MIN}, T] \approx T/2$. The only difference is that their $\Delta T_{PRBS,MAX}$ are set to T/8 and T/16 individually through the static PRBS energy-level control. The correlation function between Y_n and Y_k can be observed by plotting $Cov[Y_n, Y_k]$ of each "n" across all possible "k" (= 1 to N_{DCO}) with $N_{EXP} = 2^{13}$. For the purpose of reference, note that the correlation functions of I.I.D. scenarios (i.e., $\Delta T_{PRBS,MAX} = T$) verified in Eq. (19) are basically zero for all $n \neq k$ and have the non-zero value, 0.25 $(= P_1 \cdot P_0)$, at $n = k$. From these examples of $\Delta T_{PRBS,MAX} < T$, multiple attributes can be observed: (a) $Cov[Y_n, Y_k]$ of each "n" is self-symmetrical with respect to "k = n". (b) Because of the symmetric covariance matrix and $Cov[Y_n, Y_k] = Var[Y_n] = P_1 \cdot P_0$ when $k = n$, all correlation functions have the same profile but shift along the k-axis based on the value of "n", and therefore $Cov[Y_n, Y_k] = Cov[Y_{N_{DCO}-n+1}, Y_{N_{DCO}-k+1}]$. (c) The correlation is enhanced with decreasing $\Delta T_{PRBS,MAX}$. (d) The correlation degrades with increasing $|k - n|$. (e) More importantly, the $Cov[Y_n, Y_k]$ of each "n" has a consistent sign-alternation pattern within each correlation-envelop period which guarantees the cancelation in the total covariance sum, i.e., effective VR, though the amplitude of the correlation-envelop attenuates with $|k - n|$. In sum, the theoretical variance calculation can be greatly simplified by grouping the effect of covariance cancelation into single or multiple least-common-multiple periods (LCMP) of the sign-alternation patterns and correlation-envelop periods instead of considering all the covariance elements.

The effect of this SAVR technique can be more comprehensively demonstrated by the single-dimensional covariance sums and the overall (two-dimensional) covariance sum $Var[\overline{Y}_{SA}]$. The single-dimensional covariance sum is defined as the summation of all $Cov[Y_n, Y_k]$ from k = 1 to N_{DCO} with respect to each "n". The single-dimensional covariance sums of the I.I.D. examples are always 0.25 for all "n" since each "n" only has its non-zero covariance $Var[Y_n] = 0.25$ at $n = k$. On the other hand, the single-dimensional covariance sums of the two correlated examples ($\Delta T_{PRBS,MAX} = T/8$ and T/16) are all smaller than 0.25 across all "n" because the periodical sign-alternation patterns of their correlation functions are canceling the power of $Var[Y_n] = 0.25$ at $n = k$. This is the reason for naming this technique as self-antithetic VR. Finally, the overall variance of RSA can be obtained by the summation of all single-dimensional covariance sums divided by N_{DCO}^2 as shown at the end of 1st line of Eq. (20). The overall variances, $Var[\overline{Y}_{SA}]$, are reduced from the value of the I.I.D. variance. Equivalently, the first term in the 3rd line of Eq. (20), i.e., $P_1 \cdot P_0/N_{DCO}$, equals the I.I.D. variance without any SAVR, and the second term in the 3rd line of Eq. (20) is negative to successfully perform SAVR.

The analysis results so far for SAVR offer further insights: (a) A smaller $\Delta T_{PRBS,MAX}$, i.e., a narrower fundamental sampling PDF, $f_1(t)$, creates stronger correlations across all sampling points (i.e., longer correlation tails), longer LCMPs, and eventually more variance reduction. (b) Though the sign of each single-dimensional covariance sum can be either positive or negative with VR, the average of all single-dimensional covariance sums is always above zero, which verifies the principles of non-negative variances and finite measurement resolutions. (c) More importantly, the computation efficiency of each theoretical single-dimensional covariance sum and overall variance can be greatly improved by only summing the covariances surrounding the "k = n" within a few LCMPs [10]:

$$
\begin{gathered}
If\, \Delta T_{PRBS,MAX} = \frac{2 \bullet T}{LCMP} \; \& \; \frac{r \bullet LCMP + 2}{2} \leq n \leq \frac{2 \bullet N_{DCO} - r \bullet LCMP + 2}{2} \\
E\left[\sum_{k=1}^{N_{DCO}} Cov[Y_n, Y_k]\right] \approx \sum_{k=n-\frac{r \bullet LCMP}{2}}^{n+\frac{r \bullet LCMP}{2}-1} Cov[Y_n, Y_k]
\end{gathered} \tag{25}
$$

where "r" is the number of LCMP included in the approximation of Eq. (25). Although the accurate one-dimensional covariance sums and partial one-dimensional covariance sums formulated by the right-hand side of Eq. (25) with r = 1 and k = (n − LCMP/2) to (n + LCMP/2–1) at each "n" contain certain amounts of delta, the distribution of the partial one-dimensional covariance sums is relatively constant and basically the average of the accurate one-dimensional covariance sums except the values of "n" approach 1 or N_{DCO} since the correlation functions of these "n" have significantly unbalanced correlation-tail lengths. Therefore, any partial single-dimensional covariance sum at any "n" far away from 1 or N_{DCO} is sufficient for approximating the average of the accurate one-dimensional covariance sums formulated by the left-hand-side of Eq. (25). For instance, by plugging Eq. (25) into the end of the 1st line of Eq. (20) with "n = $N_{DCO}/2$", the overall variance can be approximated as follows [10]:

$$
Var\left[\overline{Y}\right] = \frac{E\left[\sum_{k=1}^{N_{DCO}} Cov[Y_n, Y_k]\right]}{N_{DCO}} \approx \frac{1}{N_{DCO}} \bullet \sum_{k=\frac{N_{DCO} - r \bullet LCMP}{2}}^{\frac{N_{DCO} + r \bullet LCMP}{2} - 1} Cov\left[Y_{\left(\frac{N_{DCO}}{2}\right)}, Y_k\right] \tag{26}
$$

Note that the $Cov[Y_n, Y_k]$ in Eqs. (25) and (26) is referring to its theoretical definition in Eq. (24), and the approximation errors in Eqs. (25) and (26) can be improved by extending the number of LCMP, i.e., "r", included in the summation operators.

As discussed earlier, if the theoretical computation effort is dominated by the modulo-T circular convolutions, the computation efficiency improvement from

Eq. (20) to Eq. (26) can be evaluated by the ratios between their operation numbers of modulo-T circular convolutions. With the fact of the correlation tails being roughly vanished when $|k - n| > N_{DCO}/2$, the required number of modulo-T circular convolutions in Eq. (20) can be expressed in the numerator of Eq. (27). On the other hand, for Eq. (26), only the covariance elements within one LCMP need to be calculated, and the required number of modulo-T circular convolutions in Eq. (26) can be expressed in the denominator of Eq. (27) [10].

$$\frac{\sum_{k=\frac{N_{DCO}}{2}+1}^{N_{DCO}-1}\left(k-\frac{N_{DCO}}{2}\right)\bullet 2^{\left(k-\frac{N_{DCO}}{2}+1\right)}}{\sum_{k=\frac{N_{DCO}}{2}+1}^{\frac{N_{DCO}}{2}+LCMP}\left(k-\frac{N_{DCO}}{2}\right)\bullet 2^{\left(k-\frac{N_{DCO}}{2}+1\right)}} \tag{27}$$

The ratio of modulo-T circular convolution numbers between Eqs. (20) and (26) under this specific example (i.e., $\mathrm{Mod}[T_{DCO,MIN}, T] \approx T/2$ and $\Delta T_{PRBS,MAX} \approx T/8$) is about 2.18×10^{34}. This incredible "computation effort reduction" (NOT variance reduction), even just for $N_{DCO} = 2^8$ and one LCMP included, is again mainly because all pairwise covariances theoretically have to be included in Eq. (20) due to the inevitable correlations (based on the DCO phase-noise accumulation property and Convolution Theorem) among the majority of N_{DCO} samples, but actually only the average of all the covariances matters and can be approximated by summing the theoretical covariances within a few LCMPs of the central correlation function as shown in Eq. (26). In sum, the theoretical calculation results and the statistical asynchronous RSA simulation results match well, but the computation efficiency of the theoretical variance with SAVR enabled is actually quite low if relying on the equations from Eqs. (20)-(24). On the other hand, Eq. (26) brings the computation efficiency to a reasonable level but losing the accuracy. Only the Monte Carlo approach from the statistical software/lab experiments can simultaneously offer the efficiency and accuracy from the analytical point of view [10].

6. Control-variate variance reduction

The control-variate variance reduction (CVVR) [3] employs the information of errors in estimates of pre-set quantities to reduce the variance in an estimate of an quantity under the asynchronous RSA measurement. Note that SAVR and CVVR in this chapter both reduce the power of quantization error by creating correlations among the samples of each RSA measurement. The difference is that the correlations of SAVR exist among the samples of a random variable Y, i.e., $[Y_1, Y_2, \ldots, Y_{N_{DCO}}]$ itself, which are auto-correlations [5], so the overall variance of the RSA with SAVR is quantified by an auto-covariance sum as expressed in Eq. (20); meanwhile, the correlations of CVVR exist among the samples of at least two random variables, Y and Y_{REF}, i.e., $[Y_1, Y_2, \ldots, Y_{N_{DCO}}]$ and $[Y_{REF,1}, Y_{REF,2}, \ldots, Y_{REF,N_{DCO}}]$, which are cross-correlations [5], so the overall variance of the RSA with CVVR is quantified by a cross-covariance.

The theoretical analysis of CVVR can start with the case where Y and Y_{REF} are individual I.I.D. random variables; i.e., no auto-correlation, and only cross-correlation is considered in this section. Eq. (28) shows the process of sampling $CK\tau$ and $CK\tau_{REF}$ by CK_{DCO} simultaneously to generate a cross-correlation between Y and Y_{REF} only at the sampling instants of "n = k" because Y and Y_{REF} are individual I.I.D. random variables [11]:

$$Cov[Y_n, Y_{REF,k}] = E[Y_n \cdot Y_{REF,k}] - E[Y_n] \bullet E[Y_{REF,k}] = \begin{cases} Cov[Y, Y_{REF}] \neq 0, if\ n = k \\ 0, if\ n \neq k \end{cases} \tag{28}$$

In Eq. (28), both "n" and "k" = 1 to N_{DCO}. Furthermore, the cross-covariance between the outcomes of the parallel asynchronous RSA processes, $\overline{Y}$ and $\overline{Y}_{REF}$, can be expressed as follows [11]:

$$\begin{aligned} Cov\left[\overline{Y}, \overline{Y}_{REF}\right] &= E\left[\overline{Y} \bullet \overline{Y}_{REF}\right] - E\left[\overline{Y}\right] \bullet E\left[\overline{Y}_{REF}\right] \\ &= \frac{\sum_{n=1}^{N_{DCO}} E[Y_n \bullet Y_{REF,n}]}{N_{DCO}^2} + \frac{2 \bullet \sum_{n=2}^{N_{DCO}} \sum_{k=1}^{n-1} E[Y_n \bullet Y_{REF,k}]}{N_{DCO}^2} \\ &\quad - E[Y] \cdot E[Y_{REF}] \end{aligned} \tag{29}$$

Since Y and Y_{REF} are individual I.I.D. random variables, the cross-covariance can be simplified based on Eq. (28) as follows [11]:

$$\begin{aligned} Cov\left[\overline{Y}, \overline{Y}_{REF}\right] &= \frac{N_{DCO} \bullet E[Y \bullet Y_{REF}]}{N_{DCO}^2} + \frac{2 \bullet \sum_{n=2}^{N_{DCO}} \sum_{k=1}^{n-1} E[Y_n] \bullet E[Y_{REF,k}]}{N_{DCO}^2} - E[Y] \cdot E[Y_{REF}] \\ &= \frac{E[Y \bullet Y_{REF}] - E[Y] \bullet E[Y_{REF}]}{N_{DCO}} = \frac{Cov[Y, Y_{REF}]}{N_{DCO}} \end{aligned} \tag{30}$$

As shown in Eq. (30), sampling $CK\tau$ and $CK\tau_{REF}$ by CK_{DCO} simultaneously can create a cross-correlation between $\overline{Y}$ and $\overline{Y}_{REF}$. Because of the same rising edges, clock period, and random sampling PDFs, $f_n(t)$, the only difference between the waveforms of $CK\tau$ and $CK\tau_{REF}$ is their duty-cycles. This duty-cycle delta is the factor causing the different outcomes between the data sequences of Y and Y_{REF} without considering any mismatch between the sampling DFFs. By plugging the expectations of Y, Y_{REF}, and $(Y \cdot Y_{REF})$ into Eq. (30), the cross-covariance becomes [11]:

$$\begin{aligned} Cov\left[\overline{Y}, \overline{Y}_{REF}\right] &= \frac{Cov[Y, Y_{REF}]}{N_{DCO}} = \frac{\sigma_Y \bullet \sigma_{Y_{REF}} \bullet \rho_{Y,Y_{REF}}}{N_{DCO}} = \sigma_{\overline{Y}} \bullet \sigma_{\overline{Y}_{REF}} \bullet \rho_{\overline{Y},\overline{Y}_{REF}} \\ &= \begin{cases} \dfrac{P_1 - P_1 \bullet P_{1,REF}}{N_{DCO}} = \dfrac{P_1 \bullet P_{0,REF}}{N_{DCO}}, if\ \tau \leq \tau_{REF} \\ \dfrac{P_{1,REF} - P_1 \bullet P_{1,REF}}{N_{DCO}} = \dfrac{P_0 \bullet P_{1,REF}}{N_{DCO}}, if\ \tau > \tau_{REF} \end{cases} \end{aligned} \tag{31}$$

where σ_Y, $\sigma_{Y_{REF}}$, $\sigma_{\overline{Y}}$, and $\sigma_{\overline{Y}_{REF}}$ are standard deviations of Y, Y_{REF}, $\overline{Y}$, and $\overline{Y}_{REF}$, respectively; $\rho_{Y,Y_{REF}}$ is the correlation coefficient [4] between Y and Y_{REF}, and $\rho_{\overline{Y},\overline{Y}_{REF}}$ between $\overline{Y}$ and $\overline{Y}_{REF}$. Since Y and Y_{REF} are individually I.I.D., their standard deviations are the square roots of their variances on the I.I.D. condition shown in Eq. (20). Therefore, the correlation coefficients can be found as follows based on Eqs. (20) and (31) [11]:

$$\rho_{\overline{Y},\overline{Y}_{REF}} = \rho_{Y,Y_{REF}} = \begin{cases} \sqrt{\dfrac{P_1 \bullet P_{0,REF}}{P_0 \bullet P_{1,REF}}}, \text{if } \tau \leq \tau_{REF} \\ \sqrt{\dfrac{P_0 \bullet P_{1,REF}}{P_1 \bullet P_{0,REF}}}, \text{if } \tau > \tau_{REF} \end{cases} \tag{32}$$

According to Eq. (32), the correlation coefficients are functions of τ and τ_{REF} since $P_1 = 1 - P_0 = \tau/T$ and $P_{1,REF} = 1 - P_{0,REF} = \tau_{REF}/T$ according to Eq. (6), so the degree of the cross-correlation between Y and Y_{REF} is determined by the amount of overlap between the waveforms of the CKτ and CKτ_{REF}, which is proven by Eq. (32).

Once confirming the cross-correlation between Y and Y_{REF} can be implemented by the random sampling process, the Monte Carlo estimate of Y_{REF} is obtained by the regular asynchronous RSA process to measure the duty-cycle of CKτ_{REF} (τ_{REF}/T) with a relatively high accuracy so that $\hat{E}[\overline{Y}_{REF}]$ with Var[$\overline{Y}_{REF}$] becomes pre-set and "known" information before the realistic measurement of "unknown" random variable Y with a targeted accuracy ($\overline{Y}$ with Var[$\overline{Y}$]). Because obtaining $\hat{E}[\overline{Y}_{REF}]$ with Var [$\overline{Y}_{REF}$] has to be done before measuring $\overline{Y}$ with Var[$\overline{Y}$], the time consumption due to pre-measuring $\hat{E}[\overline{Y}_{REF}]$ is under the cost of instrument initialization and calibration procedure, which shall not be counted as a part of time-consumption under each RSA-based TDC process. This is essentially how CVVR can improve the conversion-rate. Note that, instead of $\overline{Y}_{REF}$, the Monte Carlo estimate of Y_{REF} is denoted by $\hat{E}[\overline{Y}_{REF}]$ to represent its extremely high accuracy obtained from the pre-measuring process.

With the information of $\hat{E}[\overline{Y}_{REF}]$ and the cross-correlation between Y and Y_{REF}, the samples of a new random variable Y_{CV} having the identical expectation of Y but a lower variance because of CVVR can be expressed as follows based on the relation between $Y_{REF,n}$ and Y_n [3]:

$$Y_{CV,n} = Y_n - \mu_{CV} \bullet \left(Y_{REF,n} - \hat{E}\left[\overline{Y}_{REF}\right]\right) \tag{33}$$

$$E[Y_{CV}] = E[Y] - \mu_{CV} \bullet E[Y_{REF}] + \mu_{CV} \bullet E\left[\hat{E}\left[\overline{Y}_{REF}\right]\right] = E[Y] = \frac{\tau}{T} = P_1 \tag{34}$$

$$\begin{aligned} Var[Y_{CV}] &= Var\left[Y - \mu_{CV} \bullet \left(Y_{REF} - \hat{E}\left[\overline{Y}_{REF}\right]\right)\right] \\ &\approx Var[Y - \mu_{CV} \bullet Y_{REF} + \mu_{CV} \bullet E[Y_{REF}]] \\ &= Var[Y] + \mu_{CV}^2 \bullet Var[Y_{REF}] - 2 \bullet \mu_{CV} \bullet Cov[Y, Y_{REF}] \end{aligned} \tag{35}$$

where $Y_{CV,n}$ is the variance-reduced version of Y_n per sample; the error term, ($Y_{REF,n}$ - $\hat{E}[\overline{Y}_{REF}]$), serves as a "control" per sample to produce $Y_{CV,n}$ in estimating E [Y] = E[Y_{CV}] as shown in Eq. (34); μ_{CV} represents the VR coefficient. To comprehend how the effective VR can be reached by Eq. (33), the fully correlated situation ($Y_n = Y_{REF,n}$ for all "n") between Y and Y_{REF} can be quickly examined; i.e., when $\mu_{CV} = 1$, Eq. (33) leads to $Y_{CV,n} = \hat{E}[\overline{Y}_{REF}]$ for all "n", so even a single sampled outcome $Y_{CV,1}$ can reach the accuracy of $\hat{E}[\overline{Y}_{REF}]$ without a large number of samples. The degree of correlation between Y and Y_{REF} significantly affects the amount of VR and accordingly reduce the number of the required samples N_{DCO} to reach the target accuracy, and note that the achievable accuracy is dominated by Var[$\overline{Y}_{REF}$] during the pre-measuring process. That is, $Var[Y_{CV}]$ is smaller than $Var[Y]$ because the summation of the 2nd and 3rd terms at the end of Eq. (35) is negative and a function of $Cov[Y, Y_{REF}]$. In sum, the benefit of CVVR mainly relies on the pre-measuring process: (a) If the pre-measuring process offers more time or accuracy, the more VR or conversion-rate can be reached. (b) If Y_{REF} and Y have a higher correlation, the more VR or conversion-rate can be obtained as well.

There are two major concerns in the IC implementation: additional circuit power/area consumption and achievable degree of correlation. Actually, realizing $Y_{CV,n}$ in Eq. (33) requires a high speed/resolution digital operation per sample; this can consume a certain amount of power/area because of two reasons: (a) The pre-measured $\hat{E}[\overline{Y}_{REF}]$ and μ_{CV} both possess high resolutions to meet the targeted accuracy per RSA process. (b) Though $\hat{E}[\overline{Y}_{REF}]$ and μ_{CV} are constant, the digital operation of Eq. (33) is executed in the speed of the DCO frequency to produce the high-resolution $Y_{CV,n}$ per sample. The solution for these two concerns is based on the fact that the final outcome of each RSA process is the mean value of the sampled random variable $\overline{Y}_{CV}$. Therefore, the averaging process can be incorporated into Eq. (33) to perform the RSA with CVVR technique [11]:

$$\overline{Y}_{CV} = \frac{\sum_{n=1}^{N_{DCO}} \left[Y_n - \mu_{CV} \bullet \left(Y_{REF,n} - \hat{E}\left[\overline{Y}_{REF}\right]\right)\right]}{N_{DCO}} = \overline{Y} - \mu_{CV} \bullet \left(\overline{Y}_{REF} - \hat{E}\left[\overline{Y}_{REF}\right]\right)$$
$$= \frac{N_Y}{N_{DCO}} - \mu_{CV} \bullet \left(\frac{N_{Y_{REF}}}{N_{DCO}} - \hat{E}\left[\overline{Y}_{REF}\right]\right) \tag{36}$$

$$E\left[\overline{Y}_{CV}\right] = E\left[\overline{Y}\right] - \mu_{CV} \bullet E\left[\overline{Y}_{REF}\right] + \mu_{CV} \bullet E\left[\hat{E}\left[\overline{Y}_{REF}\right]\right]$$
$$= E\left[\overline{Y}\right] = E[Y] = \frac{\tau}{T} = P_1 \tag{37}$$

$$Var\left[\overline{Y}_{CV}\right] = Var\left[\overline{Y} - \mu_{CV} \bullet \left(\overline{Y}_{REF} - \hat{E}\left[\overline{Y}_{REF}\right]\right)\right]$$
$$\approx Var\left[\overline{Y}\right] + \mu_{CV}^2 \bullet Var\left[\overline{Y}_{REF}\right] - 2 \bullet \mu_{CV} \bullet Cov\left[\overline{Y}, \overline{Y}_{REF}\right] \tag{38}$$
$$= \frac{Var[Y]}{N_{DCO}} + \mu_{CV}^2 \bullet \frac{Var[Y_{REF}]}{N_{DCO}} - 2 \bullet \mu_{CV} \bullet \frac{Cov[Y, Y_{REF}]}{N_{DCO}}$$

Compared to Eqs. (33)-(35), respectively, Eqs. (36)-(38) represent a power/area efficient IC implementation of the RSA with CVVR technique: (a) Instead of finding every high-resolution $Y_{CV,n}$ in Eq. (33) and then averaging N_{DCO} samples, Eq. (36) shows that $\overline{Y}_{CV}$ can be obtained by a one-time high-resolution calculation after the parallel RSA processes of $\overline{Y}_{REF}$ (= $N_{Y_{REF}}/N_{DCO}$) and $\overline{Y}$ (= N_Y/N_{DCO}) are finished, so the high-speed random sampling-and-averaging circuits all stay in the 1-bit digital operations. (b) The expectation in Eq. (37) proves that the convergency of $\overline{Y}_{CV}$ is independent of the CVVR simplification from Eqs. (33)-(36). (c) The variance of the RSA with CVVR technique in Eq. (38) still follows the WLLN.

With Eq. (36) and pre-measured $\hat{E}[\overline{Y}_{REF}]$ ready, the optimal value of the VR coefficient μ_{CV}, which can minimize the variance of $\overline{Y}_{CV}$ shown in Eq. (38), is the last item for realizing the RSA with CVVR technique [11]:

$$\frac{\partial(Var[\overline{Y}_{CV}])}{\partial\mu_{CV}} = 2\bullet\mu_{CV}\bullet Var[\overline{Y}_{REF}] - 2\bullet Cov[\overline{Y}, \overline{Y}_{REF}] = 0$$

$$\mu_{CV} = \frac{Cov[\overline{Y}, \overline{Y}_{REF}]}{Var[\overline{Y}_{REF}]} = \frac{\sigma_{\overline{Y}}}{\sigma_{\overline{Y}_{REF}}}\bullet\rho_{\overline{Y},\overline{Y}_{REF}} \tag{39}$$

By plugging Eq. (39) into (38), the minimum variance of the RSA with CVVR technique is shown as follows [11]:

$$Var[\overline{Y}_{CV}] \approx Var[\overline{Y}] - \frac{Cov^2[\overline{Y}, \overline{Y}_{REF}]}{Var[\overline{Y}_{REF}]} \tag{40}$$

Clearly, the amount of variance reduction (the 2nd term of Eq. (40)) from the variance of an I.I.D. case (the 1st term of Eq. (40)) is mainly determined by the cross-correlation between $\overline{Y}$ and $\overline{Y}_{REF}$. Moreover, the conclusion shown in Eq. (40) can be further quantified by the gain of CVVR or conversion-rate, G_{CV}, of this technique, which is represented by the ratio between the RSA variances without and with CVVR [11]:

$$G_{CV} = \frac{Var[\overline{Y}]}{Var[\overline{Y}_{CV}]} = \frac{Var[\overline{Y}]\bullet Var[\overline{Y}_{REF}]}{Var[\overline{Y}]\bullet Var[\overline{Y}_{REF}] - Cov^2[\overline{Y}, \overline{Y}_{REF}]}$$

$$= \frac{\sigma^2_{\overline{Y}}\bullet\sigma^2_{\overline{Y}_{REF}}}{\sigma^2_{\overline{Y}}\bullet\sigma^2_{\overline{Y}_{REF}} - \sigma^2_{\overline{Y}}\bullet\sigma^2_{\overline{Y}_{REF}}\bullet\rho^2_{\overline{Y},\overline{Y}_{REF}}} = \frac{1}{1-\rho^2_{\overline{Y},\overline{Y}_{REF}}} \tag{41}$$

Based on Eqs. (39)-(41), several attributes and implementation methodologies of CVVR are discussed as follows: (a) The gain of CVVR, G_{CV}, is a function of the cross-correlation between the final outcomes of the parallel RSA processes, $\overline{Y}$ and $\overline{Y}_{REF}$. For instance, when $\rho_{\overline{Y},\overline{Y}_{REF}} = 0 = \mu_{CV}$, $G_{CV} = 1$, and then $Var[\overline{Y}_{CV}] = Var[\overline{Y}]$; when $\rho_{\overline{Y},\overline{Y}_{REF}} = 1 = \mu_{CV}$, $G_{CV} = \infty$, and then $Var[\overline{Y}_{CV}] \approx 0$ based on Eq. (40). (b) Different from the variance of SAVR in Eq. (20) requiring the auto-covariance sum of all $Cov[Y_n, Y_k]$ to be negative when $n \neq k$, the cross-covariance $Cov[\overline{Y}, \overline{Y}_{REF}]$ of CVVR can have a positive or negative sign; either way can produce effective VR instead of variance-addition (VA) since, as shown in Eq. (40), $Var[\overline{Y}_{CV}]$ is always smaller than or equal to $Var[\overline{Y}]$ regardless of the sign of $Cov[\overline{Y}, \overline{Y}_{REF}]$, which is canceled by itself embedded in μ_{CV} as shown in Eq. (39). This fact reflects the outstanding stability of CVVR than that of SAVR. (c) CVVR scales the variance inversely proportional to G_{CV} based on Eq. (41). That is, if a certain amount of VR can be reached by enlarging N_{DCO}, then the same amount of VR can be also reached by CVVR without enlarging N_{DCO}. Therefore, G_{CV} can be used to represent the gain of conversion-rate enhancement or measurement-time ($\propto N_{DCO}\cdot T_{DCO,AVG}$) reduction offered by CVVR under a certain requirement of ENOB.

The attributes mentioned above are all under the assumption of knowing optimal the value of μ_{CV}. However, based on Eqs. (30) and (39) regarding μ_{CV} and $Cov[\overline{Y}, \overline{Y}_{REF}]$, if E[Y], which is the ideal result of each RSA measurement, is unknown, then it seems unlikely to find the optimal VR coefficient. The good news is that the Monte Carlo method [3] is also applicable to finding an estimate of μ_{CV} by using the estimates from the RSA processes under a finite sampling number N_{DCO} instead of using the expectations, E[Y] and $E[Y_{REF}]$. Therefore, because Y and Y_{REF} are individual I.I.D. random variables based on the assumption of this section, the estimate VR coefficient $\hat{\mu}_{CV}$ can be obtained by [11]:

$$\mu_{CV} = \frac{Cov\left[\overline{Y}, \overline{Y}_{REF}\right]}{Var\left[\overline{Y}_{REF}\right]} = \frac{Cov[Y, Y_{REF}]}{Var[Y_{REF}]} = \frac{E[(Y - E[Y]) \bullet (Y_{REF} - E[Y_{REF}])]}{E\left[(Y_{REF} - E[Y_{REF}])^2\right]}$$

$$\approx \frac{\sum_{n=1}^{N_{DCO}} \left[\left(Y_n - \overline{Y}\right) \bullet \left(Y_{REF,n} - \overline{Y}_{REF}\right)\right]}{\sum_{n=1}^{N_{DCO}} \left(Y_{REF,n} - \overline{Y}_{REF}\right)^2} = \hat{\mu}_{CV} \tag{42}$$

Though Eq. (42) eliminates the necessity of ideal expectations, it requires hardware to store the entire data sequences of Y_n and $Y_{REF,n}$ with N_{DCO} samples until the parallel RSA process outcomes, $\overline{Y}$ and $\overline{Y}_{REF}$, are completed and ready for the operation of Eq. (42) and then Eq. (36). To minimize IC power/area consumptions, $\hat{\mu}_{CV}$ in Eq. (42) can be further derived to become Eq. (43) demonstrating significant hardware simplification in the IC implementation for finding $\hat{\mu}_{CV}$. Eq. (43) essentially performs a one-time digital operation with N_Y, $N_{Y_{REF}}$, $N_{Y \bullet Y_{REF}}$ and N_{DCO} from the 1-bit accumulators after the completion of each RSA-based TDC with CVVR technique [11].

$$\hat{\mu}_{CV} = \frac{\sum_{n=1}^{N_{DCO}} \left(Y_n \bullet Y_{REF,n} - Y_n \bullet \overline{Y}_{REF} - \overline{Y} \bullet Y_{REF,n} + \overline{Y} \bullet \overline{Y}_{REF}\right)}{\sum_{n=1}^{N_{DCO}} \left(Y_{REF,n}^2 - 2 \bullet Y_{REF,n} \bullet \overline{Y}_{REF} + \overline{Y}_{REF}^2\right)}$$

$$= \frac{\sum_{n=1}^{N_{DCO}} \left(Y_n \bullet Y_{REF,n}\right) - N_{DCO} \bullet \overline{Y} \bullet \overline{Y}_{REF}}{\sum_{n=1}^{N_{DCO}} \left(Y_{REF,n}\right) - 2 \bullet \overline{Y}_{REF} \bullet N_{DCO} \bullet \overline{Y}_{REF} + N_{DCO} \bullet \overline{Y}_{REF}^2}$$

$$= \frac{\frac{\sum_{n=1}^{N_{DCO}} \left(Y_n \bullet Y_{REF,n}\right)}{N_{DCO}} - \overline{Y} \bullet \overline{Y}_{REF}}{\overline{Y}_{REF} - \overline{Y}_{REF}^2} = \frac{N_{DCO} \bullet N_{Y \bullet Y_{REF}} - N_Y \bullet N_{Y_{REF}}}{\left(N_{DCO} - N_{Y_{REF}}\right) \bullet N_{Y_{REF}}} \tag{43}$$

7. Conclusion

The evolution of the four RSA techniques has been reviewed in this chapter, including synchronous RSA, asynchronous RSA, asynchronous RSA with SAVR, and asynchronous RSA with CVVR, whose theoretical expectations, variances, figure of merits, and parameter settings of their IC implementations are listed in **Table 1**. The main goal of this chapter is to summarize the concept/algorithm of the RSA with VR techniques and introduce the unified RSA-based TDC architecture for both high-speed and high-resolution TCSPC applications.

Technology	22-nm CMOS	22-nm CMOS	22-nm CMOS
Technique	Asyn. RSA	Asyn. RSA w/SAVR	Asyn. RSA w/ CVVR
DCO Power (mW)	3	3.1	3
Sampling Frequency (GS/s)	4	7.8	4
# of Sampling Phases	8	8	8
Dynamic Range (ns)	10 ~ 1000	10 ~ 1000	10 ~ 1000
ENOB	12 @$N_{DCO} = 2^{24}$ 14 @$N_{DCO} = 2^{28}$	12 @$N_{DCO} = 2^{19}$ 14 @$N_{DCO} = 2^{23}$	12 @$N_{DCO} = 2^{20.9}$ 14 @$N_{DCO} = 2^{24.9}$
Effective Resolution (ps @14 ENOB)	0.61 ~ 61	0.61 ~ 61	0.61 ~ 61
Conversion-Rate (kHz @12 ENOB)	2	120	16
TDC Power (mW)	1.3	1.5	1.9
TDC FoM (pJ/step @12 ENOB)	159	3.1	29
DCO + TDC Power Ratio	1	1.07	1.14
Conversion-Rate Ratio	1	60	8
TDC Area (mm^2)	0.01	0.01	0.018
Digital Filter Power (mW)	0.45	0.90	0.91
Multi-bit Digital Operations per RSA-based TDC	Eq. (6)	Eq. (6)	Eq. (36) and (43)
Theoretical Expectation	Eq. (15)	Eq. (6)	Eq. (37)
Theoretical Variance	Eq. (7)	Eq. (20)	Eq. (40)
Circuit Parameters @T = 250 ps, $T_{DCO,MIN}$ = T/2	$\Delta T_{PRBS,MAX}$ = T	$\Delta T_{PRBS,MAX}$ = T/32	$\Delta T_{PRBS,MAX}$ = T, τ_{REF} LSB = T/16

Table 1.
Comparison of RSA-based time-to-digital conversion techniques.

Author details

Tzu-Chien Hsueh
University of California, San Diego, La Jolla, USA

*Address all correspondence to: tzhsueh@eng.ucsd.edu

References

[1] Becker W. Advanced Time-Correlated Single Photon Counting Techniques, Berlin. Germany: Springer; 2005

[2] Becker W. The bh TCSPC Handbook. 7th ed. Berlin, Germany: Becker & Hickl GmbH; 2017

[3] Glasserman P. Monte Carlo Methods in Financial Engineering. New York, NY, USA: Springer; 2003. DOI: 10.1007/978-0-387-21617-1

[4] Ghahramani S. Fundamentals of Probability. Upper Saddle River, NJ, USA: Prentice-Hall; 1996

[5] Haykin S. Communication Systems. 4th ed. New York, NY, USA: Wiley; 2001

[6] Wu T, Yang R, Hsueh T-C. Random sampling-and-averaging techniques for single-photon arrival-time detections in quantum applications: theoretical analysis and realization methodology. IEEE Transactions on Circuits and Systems I: Regular Papers. 2022;**69**(4):1452-1465. DOI: 10.1109/TCSI.2021.3135833

[7] Hsueh T-C, O'Mahony F, Mansuri M, Casper B. An on-die all-digital power supply noise analyzer with enhanced spectrum measurements. IEEE Journal of Solid-State Circuits. 2015;**50**(7):1711-1721. DOI: 10.1109/JSSC.2015.2431071

[8] Oppenheim AV, Schafer RW, Buck JR. Discrete-Time Signal Processing. 2nd ed. Upper Saddle River, NJ, USA: Prentice-Hall; 1999

[9] Hsueh T-C, Balamurugan G, Jaussi J, Hyvonen S, Kennedy J, Keskin G, et al. A 25.6Gb/s differential and DDR4/GDDR5 dual-mode transmitter with digital clock calibration in 22nm CMOS. In: IEEE ISSCC Digital Technical Papers. San Francisco, CA, USA: IEEE; 2014. pp. 444-445. DOI: 10.1109/ISSCC.2014.6757506

[10] Wu T, Hsueh T-C. A high-resolution single-photon arrival-time measurement with self-antithetic variance reduction in quantum applications: Theoretical analysis and performance estimation. IEEE Transactions on Quantum Engineering. 2022;**3**:1-15. DOI: 10.1109/TQE.2022.3209211

[11] Yang R, Wu T, Hsueh T-C. A high-accuracy single-photon time-interval measurement in mega-Hz detection rates with collaborative variance reduction: theoretical analysis and realization methodology. IEEE Transactions on Circuits and Systems I: Regular Papers. 2023;**70**(1):176-189. DOI: 10.1109/TCSI.2022.3206406

Chapter 5

Path Integrals and Smart Monte Carlo – I

Jan W. Dash and Xipei Yang

Abstract

We present "Smart Monte Carlo" or SMC, improving the efficiency of Monte Carlo (MC) simulations. SMC has two "stages". The first stage, run adaptively for each deal, produces equivalent results to standard MC simulation using fewer calls to the time-consuming pricing functions. The second SMC stage is standard MC simulation for a portfolio, using price look-ups from the first SMC stage. The result has improved accuracy and speed. Examples are given.

Keywords: Smart Monte Carlo, Monte Carlo efficiency, scout paths, path integrals/ Brownian bridges, interpolated Brownian bridges, portfolio

1. Introduction

Here is Smart Monte Carlo (SMC) presented in a summary recipe format. The SMC first stage, for a given deal, produces a "look-up table" of prices via three steps:

1. "*Scout Path SP*" generation [standard MC with limited number of paths giving deal prices for some values in the underlying variable(s) over time]

2. "*Brownian Bridge BB*" generation [uses path integrals, adaptive filtration based on the SP results, providing deal prices in "interesting" regions where the deal lives]

3. *Interpolation* [other deal prices based on one and two without calling the pricing function]

The SMC second stage is a standard MC simulation that however never calls the pricing functions, instead using the deal prices from SMC first stage look-up tables.

Numerical results are encouraging and substantial, as seen in **Table 1** below.

There are two sequel SMC papers. The second paper [1] expands SMC to path dependent deals and multifactor models. It also contains two more SMC ideas for improving efficiency of simulations:

- Time interpolation of deal prices. The idea is to get prices at an intermediate time t via interpolation from neighboring times without calling the pricing function at time t.

IntechOpen

Name	Description	"Speed-up factor"
Caps	Four out-of-money caps	3
Barrier	Down-out option with rebate	2
"Logo"	2D Matlab Logo reproduction	>3

Table 1.
Results for "speed-up factors" (ratio of the number of pricing function calls for standard MC relative to SMC, for similar or less uncertainty.

- Model perturbation theory PT for both time interpolation and spatial interpolation in underlying variable(s). The idea is to interpolate a "model-model price spread", the difference between the deal price and a simple approximate model "*SAM*" price.

The third paper [2] deals with SMC applied to American options.

We believe SMC is general and practically useful. For the present prototype work, the intent is to blaze some new trails and demonstrate useful concepts, without trying to "build a highway".

2. SMC Stage 1: three steps

2.1 Motivation: improve Monte Carlo (MC) efficiency

There is a well-known problem with Monte Carlo (MC) simulations. MC simulations start with stochastic equation(s) modeling the dynamics of underlying variables (stock prices, rates etc). Here is the problem. The instruments to be priced and/or risked have discounted cash flows CFs depending on these underlying variables, but whose profiles do not line up with the MC paths very well. An example is an out-of-the-money (OTM) option for which the CFs are mainly in regions with a relatively small proportion of MC paths. To get accurate results (given a stochastic model), a large number of MC paths must be run, and the option pricing function is called for every path. This is inefficient, because the option price on many (most) paths is simply negligible. This inefficiency is the problem we address in this work on SMC. Results equivalent to MC simulation accuracy are obtained with speedup of a factor of at least 2.

2.2 Scout paths (SP)—first step

The first step is to get an idea of what the deal "looks like" by generating some standard MC paths, which we call "scout paths" or SPs. The pricing function is called for the SPs and the information is returned.

2.3 Path integrals/Brownian bridges (BB): second step

The second step is to use the SPs as information to generate more efficient sampling of the deal value. This is the core idea of our resolution to the MC inefficiency problem. Consider an OTM[1] option. We would like to send paths into the OTM region

[1] OTM means out of the money with respect to today's price.

without wasting time generating paths that are in regions where the deal value is negligible for the purposes or pricing/risk. We can achieve this goal using path integrals and Brownian Bridges (BBs).

2.3.1 What are path integrals and Brownian bridges?

An introduction to path integrals is in the appendix and discussed in depth in Ref. [3], from original work by J. W. Dash in the 1980s. Some summary features are:

- Path integrals have a simple math formalism—uses the probability-based approach consistent with the stochastic equations

- The path integral framework is very intuitive.

- The path integral can be made rigorous since these path integrals do not have to deal with oscillations as in quantum mechanics (no square root of −1).

Here, the reader only has to know that a definite probability (consistent with the stochastic equation) is obtained for all paths that start from a given point at a given time, traverse all allowed points at all intermediate times, and wind up in a given "bin" region of the variable(s), at a later time. Such an (infinite) set of paths is called a "Brownian Bridge" or BB. For an OTM option, we want send BBs into the OTM region (with respect to today's price) to sample the option in a more detailed way than ordinary MC for the same number of calls to the pricer.

Here is a picture of BBs in time, with a specific path of BBs in dark font. The times can be macroscopic, using the convolution of probabilities as ordinary integrals (**Figure 1**).

How do we know a-priori where to send the BBs, e.g. where the OTM region is? Again recall the first step that runs a limited, although not **too** limited, MC simulation of SPs for which the pricing function is called. The SPs come back with enough information about "interesting" regions where substantial deal value exists. For step 2, we send BBs to those "interesting regions" [4].[2]

For further discussion, see Ref. [1], (Appendices 1–3).

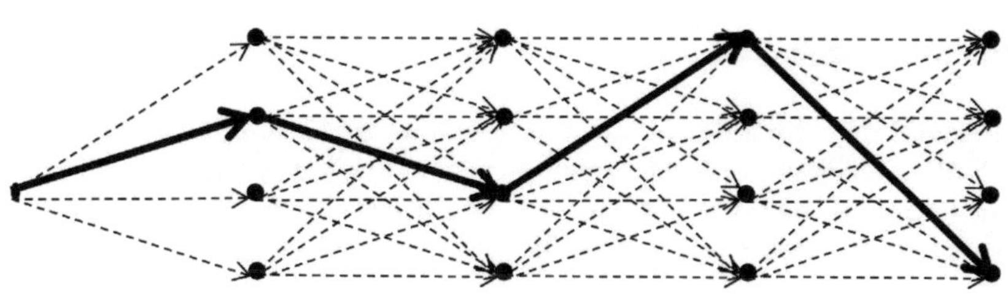

Figure 1.
Brownian bridges BBs, with one path of BBs in dark font. The times can be macroscopic.

[2] Our BB construction is conceptually similar to some Metropolis-like filter that selects MC paths according to some criteria, with the criteria here depending on the financial instrument. Note that randomness still exists with a simple filter of MC paths; the method here is not a simple filter.

2.4 Interpolated Brownian Bridges (IBB): third step

The third step is that we carry out interpolation for points sampled neither by SPs nor BBs. We term a path for which the interpolation takes place as an "interpolated Brownian Bridge" or IBB. This enables an effective "pricing" for all values of the underlying variables without calling the pricing function "too many times".

Call $\mathbf{x}^*$ a point at which we want to interpolate the deal with price C. We get $C(\mathbf{x}^*)$ from the SP and BB values: $\left\{C\left(\boldsymbol{x}_b^{(SP)}\right)\right\}, \left\{C\left(\boldsymbol{x}_b^{(BB)}\right)\right\}$. To do this we need a distance measure and an interpolation function, discussed next.

2.4.1 Distance measure and interpolation function

Denote the square of the distance from point $\mathbf{x_b}$ to point $\mathbf{x}^*$ as $d^2(\boldsymbol{x}^*, \boldsymbol{x_b}) = \|\boldsymbol{x}^* - \boldsymbol{x_b}\|^2$ for a certain number of nearest SP, BB neighbors $\mathbf{x_b}$ with respect to $\mathbf{x}^*$. Based on some analysis, we preliminarily choose five nearest neighbors as the default. As the distance gets small, the number of nearest neighbors needs to decrease in order to reproduce the value at $\mathbf{x}^*$.

We choose a nearest-neighbor cutoff "Cauchy" interpolation weight using the squared distance function $d^2(\boldsymbol{x}^*, \boldsymbol{x_b})$. The Cauchy distance bandwidth cutoff parameter κ is used to control smoothness vs. variation.[3]

$$w(\boldsymbol{x}^*, \boldsymbol{x_b}) = \frac{\left[d^2(\boldsymbol{x}^*, \boldsymbol{x_b}) + \kappa^2\right]^{-1}}{\sum_{\boldsymbol{x}_{b'}=5 \text{ nearest nbs to } \mathrm{x}^*} \left[d^2\left(\boldsymbol{x}^*, \boldsymbol{x}_b'\right) + \kappa^2\right]^{-1}} \tag{1}$$

2.4.2 Interpolated value of deal

The final interpolated value of the deal at $\mathbf{x}^*$ is then

$$C_{\text{interp}}(\boldsymbol{x}^*) = \sum_{\boldsymbol{x_b}=5 \text{ nearest nbs to } x^*} C(\boldsymbol{x_b}) w(\boldsymbol{x}^*, \boldsymbol{x_b}) \tag{2}$$

3. SMC Stage 2: running the portfolio

3.1 Setup for portfolio run

The second stage for SMC simulation is a standard MC simulation. However, this second SMC stage never calls the pricing function, but rather uses the values of deals using the output of the three steps of the first SMC stage, namely price "look-up" tables that result from the prices from the SPs, BBs, and interpolation. Both stages constitute "Smart Monte Carlo".

3.2 Full MC simulation of the SMC Stage 2

Consider a portfolio that consists of deals $\{C_l\}$. In Stage 1, we create the look-up tables for individual deals. Again, a look-up table lists the deal values produced from the

[3] We reduce the number of nearest neighbors when $\mathbf{x}^*$ is close to one of the $\mathbf{x_b}$ points. The number 5 of nearest neighbors for numerical applications here was determined empirically.

interpolation for all points on a regular grid created in the space of the risk factors. Because the interpolation is conducted in a stepwise fashion, the number of look-up tables for the prices of a given deal equals the number of time steps in the simulation. The collection of the look-up tables provides the deal values on any path that has evolved arbitrarily. It is this feature that leads to substantial time saving in Stage 2 that only looks up deal values using the tables without either calling the pricer or doing the on-the-fly interpolations. Note that Stage 2 analysis does not need to compute the SMC probabilities. In a standard MC simulation, all paths are equally weighted with a total weight of 1.

The general procedure for the Stage 2 analysis consists of the following steps:

1. Run SMC Stage 1 analysis for each deal C_l. The SMC is customized for each type of deal using the strategies described above.

2. Create the SMC Stage 1 look-up tables for each deal $\{C_l\}$.

3. Generate standard MC paths for Stage 2.

4. Look up values for each deal along each MC path, which involves looking up through a table at each time in the time partition defining the MC simulator.

3.3 Deal/risk aggregation (Stage 2)

In Stage 2, the portfolio risks (expected values and PFEs - See Appendix) are calculated by aggregation—summing up the risks of individual deals for all MC paths. For each time step, a MC path provides all the necessary information in terms of the values of the underlying risk factors that are used to determine the deal values from the SMC look-up tables. Theoretically, there is no limit on the number of Stage 2 paths except for the possible constraint concerning the time for aggregating the deals. Deals with simple structures and complicated structures are treated in the same way. More importantly, the aggregation has no offset effect on the efficiency gained in the SMC Stage 1, in which SMC paths are customized to line up with the risk profiles of individual deals.

In summary the procedure is:

1. Aggregate the values/risks of all deals (portfolio) for all MC paths.

2. Calculate the portfolio expected values and PFE, or any other desired statistic, as in a conventional MC simulation.

4. Smart MC vs. standard MC: examples

4.1 OTM caps portfolio

The first illustrative example consists of 4 out-of-the-money OTM caps.[4] This illustrates the ability of the SMC method to sample interesting regions more efficiently

[4] The caps were taken as maturing in 4 years with strikes 50, 150, 250, and 350 bp respectively. The current Libor rate was taken as 40 bp, lognormal volatility constant at 15%.

than standard MC (which only knows about the stochastic equations but not the financial instrument profile).[5] **Figure 2** shows the results.

4.1.1 Smart MC results for OTM caps portfolio

The efficiency of the SMC "new method" using 10 SPs and 20 BBs (30 calls to each pricer) is demonstrated below. **Tables 2** and **3** compare the expected values and the 90% PFEs for SMC against a conventional "regular" 100-path MC simulation (100 calls to each pricer). The uncertainty is estimated by doing the whole comparison procedure 10 times. The "true" portfolio value is defined using a large number of MC paths (10,000). See footnote 5 for more detail.

We thus cut down the number of evaluations by a factor of 3 using the SMC approach. We also get better results for values (expected and PFE) compared with the true value, along with smaller uncertainties (standard deviation).

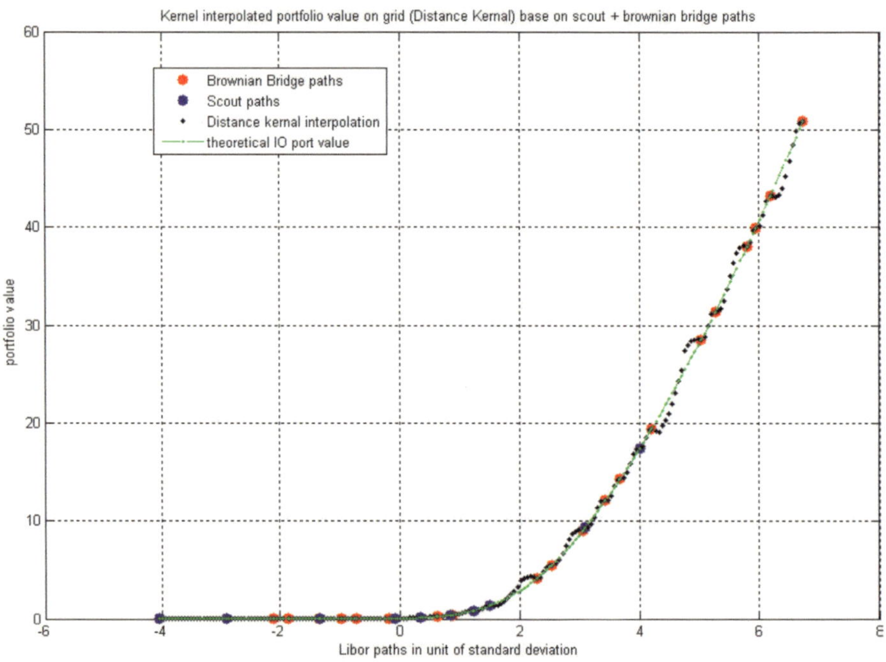

Figure 2.
Theoretical portfolio values and interpolated portfolio values from SPs (blue dots) and BBs (red dots), along with interpolated values (black dots), plotted vs. Libor in units of standard deviation, at three future times t_l (l = 1, 2, 3) *years.*

[5] In Stage 1 of SMC, we use 10 SPs (step 1) and 20 BBs (step 2). Hence we call the pricing function 30 times for each cap at each time before maturity. The positions of the BB paths are determined algorithmically to emphasize the rate region in which the portfolio of OTM caps has a "substantial" non-zero payoff. For step 3, the SPs and BBs are the basis for the interpolator that estimates the value semi-parametrically, as a function of the underlying rate. That is, for any path not sampled by either SPs or BBs, the portfolio value is approximately determined without calling the pricing function. To this end, we position 200 IBBs that partition the rate space into 200 bins around the SPs and BBs. The portfolio values over the IBBs are obtained through the interpolation, and the path integral probabilities of 200 bins are computed offline.

Expected value averaged over 10 simulations				
	No. of paths/sim	$C(t_1)$	$C(t_2)$	$C(t_3)$
Regular	100	2.1	2.1	1.2
New method	10 + 20 (intpl 200)	2.7	2.2	1.2
True	10,000	2.7	2.2	1.3
Standard deviation of expected value over 10 simulations				
	No. of paths/sim	$dC(t_1)$	$dC(t_2)$	$dC(t_3)$
Regular	100	0.14	0.14	0.14
New method	10 + 20 (intpl 200)	0.09	0.10	0.11
True	10,000	0.00	0.00	0.00

Table 2.
Expected value and error for OTM caps portfolio.

90% PFE averaged over 10 simulations				
	No. of paths/sim	$C(t_1)$	$C(t_2)$	$C(t_3)$
Regular	100	6.0	5.2	4.1
New method	10 + 20 (intpl 200)	6.6	6.5	3.9
True	10,000	6.7	6.2	3.8
Standard deviation of PFE over 10 simulations				
	No. of paths/sim	$dC(t_1)$	$dC(t_2)$	$dC(t_3)$
Regular	100	0.47	0.77	0.88
New method	10 + 20 (intpl 200)	0.25	0.35	0.46
True	10,000	0.00	0.00	0.00

Table 3.
PFE and error for OTM caps portfolio.

4.2 Down and Out DO option with rebate

To test the Smart MC (SP + BB + IBB) method in the presence of a barrier, we use the classic down and out (DO) option along with a rebate paid at touch if the barrier is crossed.[6] We get the expected value and the PFE at 90% CL.

The results are in **Tables 4** and **5**. The SMC method wins over a standard MC simulation by saving 1/2 of calls to the pricing function (50 for SMC from 30 SPs and 20 BBS, vs. 100 for standard MC) at each future time $t_\ell, l = 1...5$.

The down and out option addresses how we tackle the edge effect characterized by the difference of the option value above the barrier and the rebate below the barrier. Hence, the BBs need to be carefully positioned to separate these two states. This can

[6] The option is taken as 5 year maturity, strike 100 and barrier 80. The rebate is 5% of notional paid into a bank account that accrues interest at 1%/year. The initial price is 100 with lognormal vol 15%.

easily be achieved by placing two BBs above and below the barrier with equal distance to the barrier, such that the bins around the BBs share a same edge at the barrier. Again, we are taking advantage of the discretion feature of the BBs that can be flexibly customized to either emphasize the interesting region or better classify the states.[7]

4.2.1 Smart MC results for DO option

Expected value averaged over 10 simulations						
	No. of paths	$C(t_1)$	$C(t_2)$	$C(t_3)$	$C(t_4)$	$C(t_5)$
Regular	100	15.1	14.0	14.6	14.0	13.5
New method	30 + 20 (intpl 200)	16.4	15.7	15.2	14.9	14.6
True	10,000	16.3	16.2	16.2	15.6	15.7
Standard deviation of expected value over 10 simulations						
	No. paths/sim	$dC(t_1)$	$dC(t_2)$	$dC(t_3)$	$dC(t_4)$	$dC(t_5)$
Regular	100	0.34	0.69	1.24	1.19	1.36
New method	30 + 20 (intpl 200)	0.08	0.04	0.02	0.03	0.05
True	10,000	0.00	0.00	0.00	0.00	0.00

Table 4.
Expected value and error for down and out option.

90% PFE averaged over 10 simulations						
	No. of paths/sim	$C(t_1)$	$C(t_2)$	$C(t_3)$	$C(t_4)$	$C(t_5)$
Regular	100	27.0	27.7	38.0	39.8	45.6
New method	30 + 20 (intpl 200)	29.6	36.2	41.4	46.9	50.9
True	10,000	28.6	34.9	39.9	43.6	51.3
Standard deviation of PFE over 10 simulations						
	No. paths/sim	$dC(t_1)$	$dC(t_2)$	$dC(t_3)$	$dC(t_4)$	$dC(t_5)$
Regular	100	1.28	0.95	2.65	2.65	7.81
New method	30 + 20 (intpl 200)	0.29	0.52	0.65	0.56	0.49
True	10,000	0.00	0.00	0.00	0.00	0.00

Table 5.
PFE and error for down and out option.

[7] The probabilities attached to bins are computed differently for different states. For the bins above the barrier, the path integral probability is given by the product of the non-knock-out Green function (Ref. [3], Ch. 17) and the bin size in logarithmic scale. However, the bins below the barrier contain the paths being alive up to the previous step and knocked out at the current step. Hence, at step n, the probability is given by the product of the non-knock-out probability up to time step $n-1$ and the knock-out probability over the time interval steps $[n-1, n]$. This computation can be done either numerically or analytically. The former requires a numerical integration over the non-knock-out range at time step $n-1$, which we used here.

4.3 Reconstruct the Matlab logo—2 dimensional example

In this section, we demonstrate the efficiency of the interpolation with a complicated "payout" function in two dimensions using the Matlab [5] Logo. The logo is the drawing of the leading eigenfunction of the wave equation with a "Pac-man" spatial geometry from the Matlab logo function, in **Figure 3**. Financial payoffs generally are less complicated, so this is a good test.

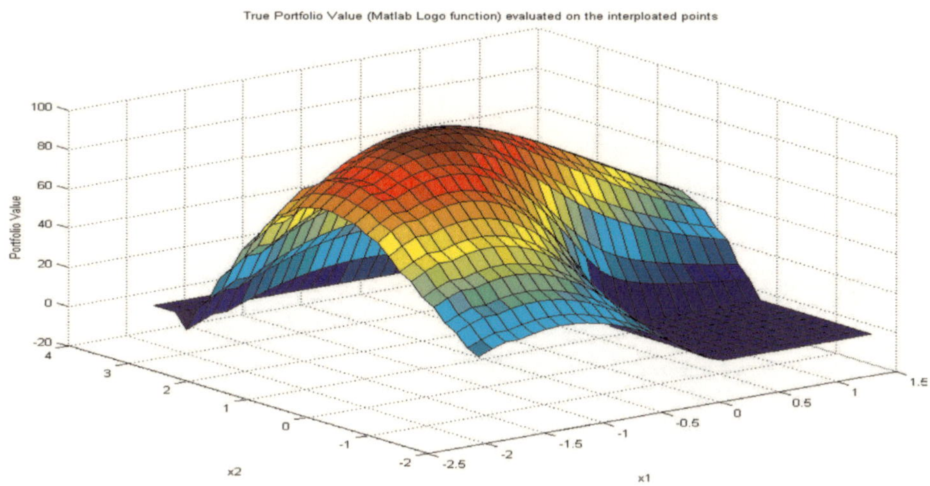

Figure 3.
The Matlab logo.

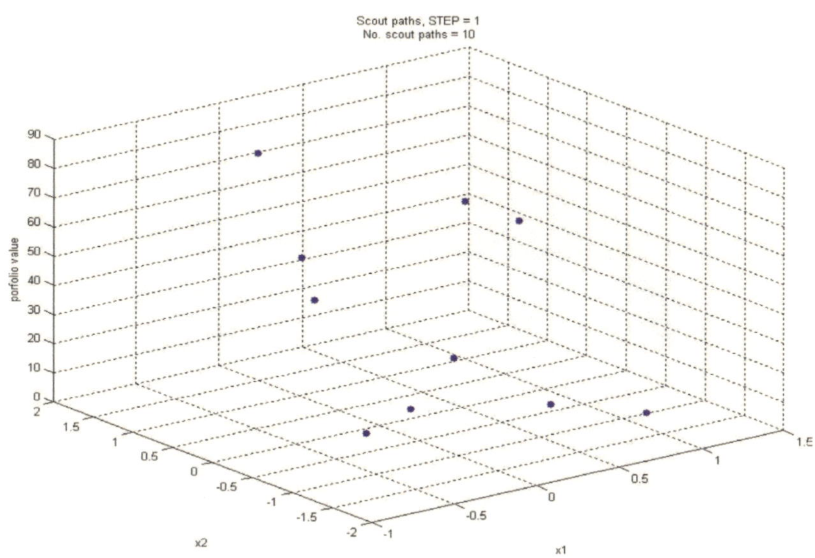

Figure 4.
The Matlab logo for the 10 SPs in SMC.

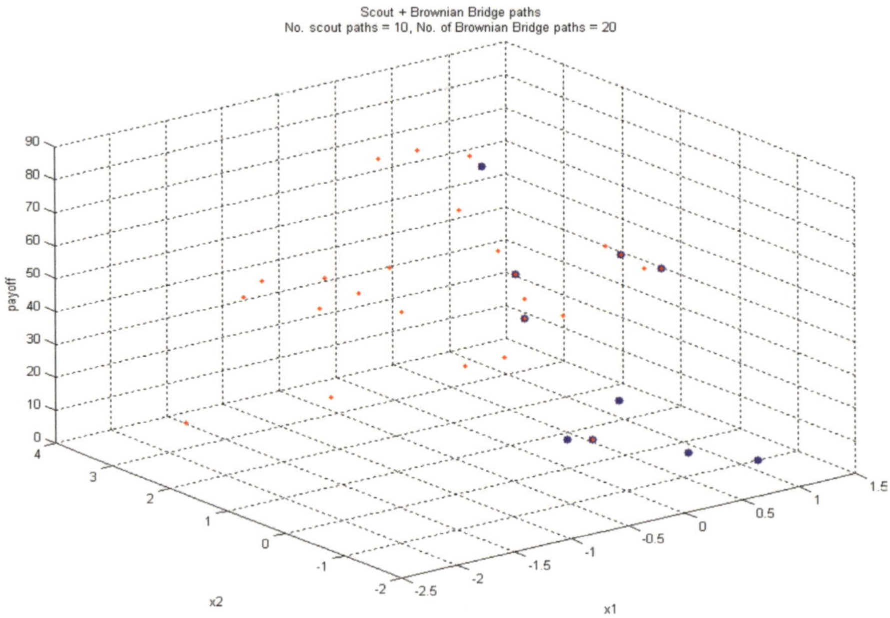

Figure 5.
The Matlab logo for the 20 BBs (red) and 10 SPs (Blue).

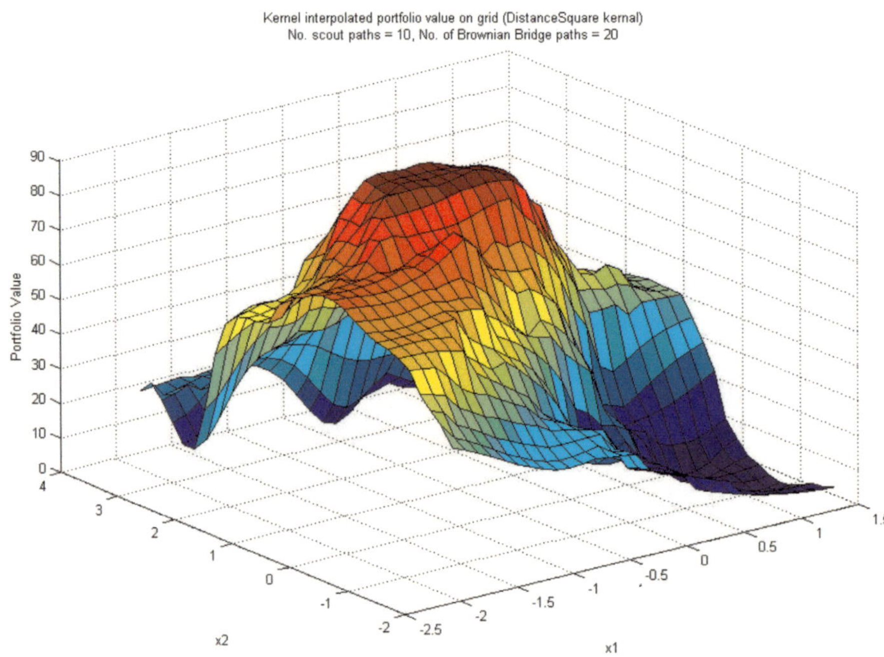

Figure 6.
The SMC interpolated result for the Matlab logo with the SP and BB input.

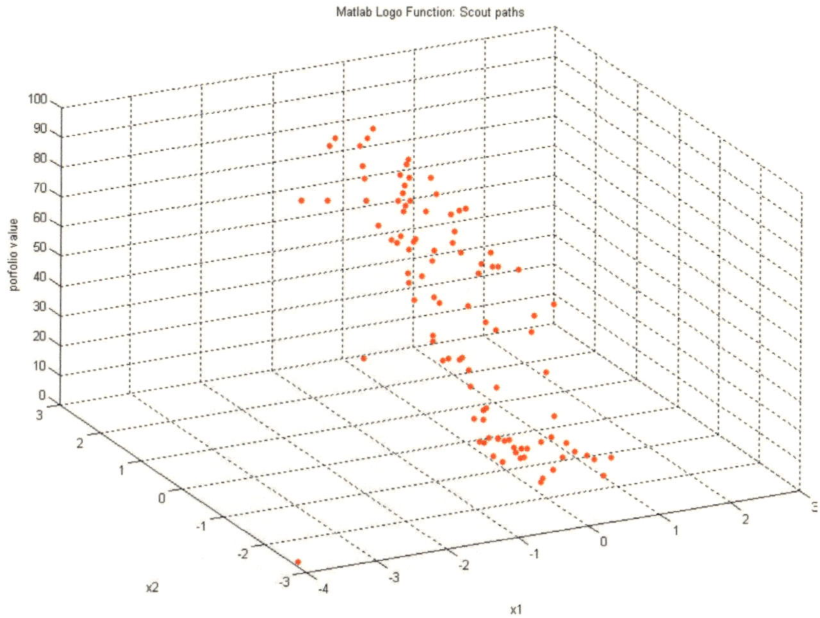

Figure 7.
Standard MC result for the Matlab logo with 100 paths.

Figures 4–6 illustrate the stepwise results in the new method.[8] We use 10 SPs (**Figure 4**) followed by 20 BBs (**Figure 5**). Here the Matlab logo from a "path" is really just a call to the Matlab logo function. With only 30 points in the 2 dimensional space calling the "pricing function" (i.e. the logo function) little resemblance is visible. Still, the interpolated result using the Cauchy distance kernel with five nearest neighbor points is fairly accurate (**Figure 6**). The SMC method Stage 1 with 30 calls to the Matlab logo function in the three steps gives a better reproduction of the Matlab logo than do the standard MC SPs with 100 calls to the Matlab logo function in **Figure** 7, a savings of at least a factor 3.

Appendix: expected value and PFE with SMC

The expected value of a portfolio is the probability weighted average of the portfolio prices. In a standard MC simulation, the expected value is the same as the sample mean of the deal prices on the simulated paths which are generated based on the underlying variable distribution assumptions. However, in the path integral simulation, the expected value is the sum of the "interpolated Brownian Bridge" values multiplied by the corresponding bin probabilities. Here, for a given bin, the probability is the product of the path integral Green function from the initial point to the center of that bin, multiplied by the size of the bin (Ref. [3], Ch. 44). This is just a Riemann approximation to an ordinary calculus integral. We use a simple illustrative portfolio; the details are not important.

The red line in **Figure 8** indicates the position of the expected value. The potential future exposure (PFE) in green is defined as the undiscounted portfolio value at a

[8] The "portfolio value" vertical axis is the height of the Matlab logo function above each point (x1, x2) in the plane.

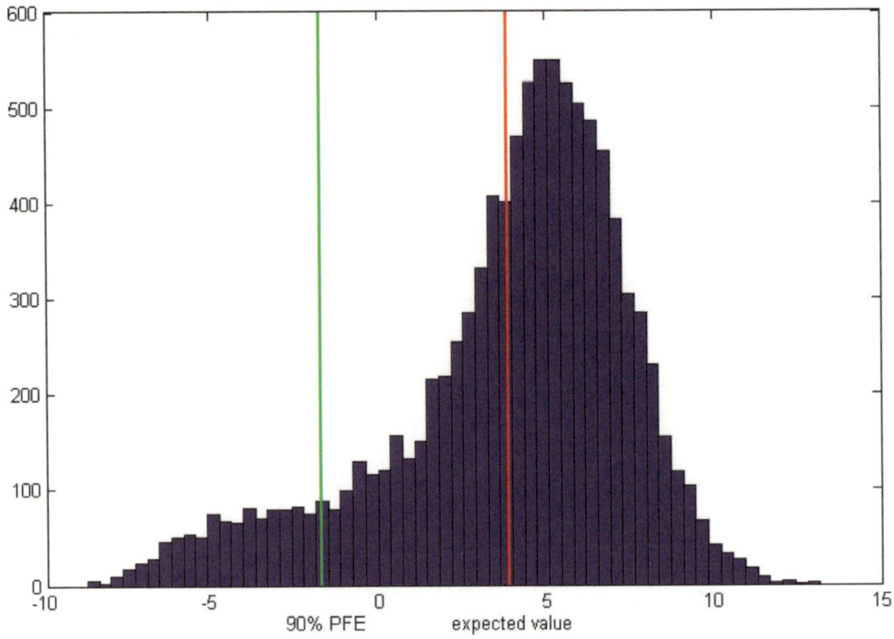

Figure 8.
Distribution (vertical axis) of simulated future illustrative portfolio values (horizontal axis); expected value (red), 90% CL PFE value (green).

given future time calculated at some level of confidence, e.g. 90%. In the SMC simulation, the PFE procedure is[9]:

- Rank the IBBs for portfolio values from smallest to largest
- Compute the empirical cumulative distribution function CDF for the IBBs by sequentially adding the bin probabilities for the value-ordered IBBs
- The 90% PFE is the 10% = 1–90% percentile of the empirical CDF.

Note

This paper was first published July, 2016 at SSRN (paper 2808177): https://ssrn.com/abstract=2808177 or http://dx.doi.org/10.2139/ssrn.2808177. Republished with permission from Bloomberg L.P. This work was created when the authors were employed at Bloomberg L.P.

[9] PFE measures the counterparty credit risk in the future during the lifetime of existing trades. In statistical terms, the 90% PFE is the 90% percentile of the distribution of forward portfolio values. For example in standard MC with 100 paths, the 90% PFE is given by the 10th smallest portfolio value since all paths are equally weighted.

Author details

Jan W. Dash[1,2]* and Xipei Yang[3]

1 Gabelli School of Business, Fordham University, New York, United States

2 J. Dash Consultants LLC, Maryland, United States

3 Citadel LLC, New York, United States

*Address all correspondence to: jdash9@comcast.net

References

[1] Dash JW, Yang X. Path Integrals and Smart Monte Carlo - II, Bloomberg L.P. working paper, 2016 (SSRN paper 2808179)

[2] Dash JW, Yang X. Smart Monte Carlo, path integrals, and American options, Bloomberg L.P. working paper, 2016 (SSRN paper 2808186)

[3] Dash JW. Quantitative Finance and Risk Management, A Physicist's Approach. 2nd ed. Singapore: World Scientific Publishing Co. Pte. Ltd.; 2016. ISBN 978-981-4571-23-4. See chapters 41–45

[4] Atchadé YF, Rosenthal JS. On adaptive Markov chain Monte Carlo algorithms. Bernoulli. 2005;**11**:815-828

[5] Matlab, copyright The MathWorks, Inc.